ENERGY FROM SOLID WASTE

ENERGY FROM SOLID WASTE

Frederick R. Jackson

NOYES DATA CORPORATION
Park Ridge, New Jersey   London, England
1974

Copyright © 1974 by Noyes Data Corporation
  No part of this book may be reproduced in any form
  without permission in writing from the Publisher.
Library of Congress Catalog Card Number: 74-75004
ISBN: 0-8155-0528-0
Printed in the United States

Published in the United States of America by
Noyes Data Corporation
Noyes Building, Park Ridge, New Jersey 07656

Foreword

It has gradually become apparent in the past few years that the most important method of disposing of solid wastes in the U.S., i.e., sanitary landfill, can no longer continue at the same rate, due to lack of land, particularly near large urban centers. An alternate disposal technique, incineration, is practiced on a much smaller scale, due to high capital and operating costs as compared to sanitary landfill. The air pollution problem created by incinerators can be solved quite easily with known technology. Solid wastes can be considered a low-sulfur fuel, consequently there is no sulfur dioxide removal problem.

There have been some installations in the United States whereby a waste heat boiler has merely been installed in a conventional incinerator. However, this method is quite inefficient as compared to a planned system whereby the incinerator and the steam producing unit are integrated with proper engineering and design considerations. There has been considerably more experience in Europe on the combustion of solid wastes to create steam, due to lack of land and higher fuel costs.

Recently, many municipalities in the United States have been investigating the feasibility of burning solid wastes to create steam for heating or electric power generation. Studies have been financed by the Environmental Protection Agency. The St. Louis plant is already on-stream and the Nashville plant will be operational in early 1974.

The energy crisis that descended upon the world in the Fall of 1973 has added a new emphasis to this waste disposal problem. It is not only the potential shortfall in supply of petroleum-based fuels that is creating the feverish new look at techniques to remove energy from solid wastes, but the cost of fuel is rising so rapidly that the combustion of solid wastes to create energy is now beginning to make economic sense.

An estimate currently making the rounds in the energy and investment communities is that when the price of crude oil reaches $7.00 per barrel, alternate sources of energy become economically viable. At the time of this writing, the U.S. controlled price of previously discovered crude oil is already $5.25 per barrel, and the price of newly discovered oil is over $8.00 per barrel. Persian Gulf crude oil is now over $7.00 per barrel (the $11.65 posted price is not an "actual" price), and adding the costs of transporting it to the United States, its actual landed value is about $9.00 per barrel.

Solid wastes, contrary to conclusions based on a superficial look, are not without their own costs. In any event they must be collected and brought to a central area for disposal by one of the known techniques. However, before being fired under a boiler, they must be processed: magnetic materials removed, shredded, conveyed, stored, etc.; and the ash removed and disposed of.

This book is based upon information available at the end of 1973, primarily from studies conducted under the auspices of the EPA. It is mainly concerned with the burning of solid wastes to create steam directly. Examples are shown based on studies for both large and small cities. Another technique that may assume more importance in the future is the production of pyrolysis gas or oil from solid wastes, and a chapter is devoted to that topic. Chapter 7 discusses European practice, which is historically far more extensive than that of the United States. The final chapter gives additional information relating to specific processes.

The Table of Contents is designed in such a way as to serve as a Subject Index.

Contents and Subject Index

INTRODUCTION 1
  Waste Heat Recovery 1
  Waste Processing 3
  Pyrolysis Products 3
REFUSE AS A LOW SULFUR FUEL 6
  Refuse As a Fuel 6
    Quantity-Quality Considerations 6
    Reduction in Pollutant Emissions 7
    Thermal Utilization 8
  Power Plant Designs 9
    New Plants 11
    Retrofit Plants 12
  Energy Utilization 14
    Turboelectric Generation 14
    Other Applications 15
ST. LOUIS PROJECT 19
  Recycling Waste As Supplementary Fuel for Power Plants 19
    Background 20
    Process Details 21
    Processing Facilities 22
    Receiving and Firing Facilities 24
    Test Boiler 25
    Potential Boiler Operating Problems 26
    Economics 27
    Applicability of Process 27
LYNN, MASS.: REFUSE-FIRED WATERWALL INCINERATOR STUDY ... 29
  Basic Design Criteria 30
  Basic Description 33

Units of Refuse Processing System 36
Units of Refuse Burning System 40
Special Considerations 44
Costs 46
Commercial Development 46

PHILADELPHIA, PA.: FEASIBILITY STUDY 47
Waste Management Operations 47
    Refuse Inventory and Composition Projections 49
    Fuel Characteristics of Philadelphia Refuse 55
    Utility Steam Generation Operations 58
Preliminary Planning Recommendations 60
    Overview 60
    District Steam Plant 61
    Power Plant 69

CLEVELAND, OHIO: FEASIBILITY STUDY 75
Waste Management Operations 75
    Refuse Inventory and Composition Projections 80
    Fuel Characteristics of Cleveland Refuse 83
    Utility Steam Generation Operations 87
Preliminary Planning Recommendations 88
    Overview 88
    Refuse-Fired District Heating Plant 89
    Refuse-Fired Turboelectric Plant 97
    Transportation Costs 101
    Conclusions 104

PYROLYSIS PROCESSES 106
Fluidized Bed Pyrolysis 106
    Process Development 108
    Process Description 112
    Experimental Gas Production Results 114
    400 Ton per Day Facility 116
    Utilization of Pyrolysis Gas 118
    Status 119
Other Pyrolysis Processes 120

EUROPEAN PRACTICE 125
Overview 125
Description of Specific German Plants 126
    Munich North, Block I 126
    Munich North, Block II 127
    Dusseldorf 127
    Stuttgart 127
    Data Analysis 128
Other Refuse-Fired Plants 128
    Essen-Karnap 128
    Berlin-Ruhleben 132
    Munich-South 132

Mannheim 133
Frankfurt am Main 134
Issy-les-Moulineaux 134
Ivry 135
Emission Control Equipment 135
Gaseous Emissions 135
PROCESS DESCRIPTIONS 136
Horner and Shifrin Fuel Recovery Process 136
Kinney Thermal Recovery System 139
Combustion Power Company Process 141
American Thermogen High Temperature Process 141
Torrax High Temperature Incinerator 146
Chicago Northwest Incinerator 148
Montreal Incinerator 151
USBM Hydrogenation Process 153
Garrett Pyrolysis Process 155
Union Carbide Oxygen Refuse Converter System 157
Monsanto Landgard System 157
Bureau of Mines Pyrolysis Process 161
Battelle Pyrolysis Incineration Process 161

Introduction

The solid waste disposal problem in the United States is reaching alarming proportions. The United States currently produces close to 300 million tons of solid wastes per year, which is expected to increase to over 340 million tons by 1980. This is equivalent to about 1 ton of solid waste per person per year. The most important methods of disposal are dumping and sanitary landfill. These disposal techniques will be difficult to continue, particularly in populated areas. Municipal incineration disposes of a small portion of our solid wastes, with attendant air and water pollution, as well as high capital and operating costs.

Many methods have been proposed for coping with the problem, such as source reduction, source separation, and material recovery. However, with the energy crisis predicted to be with us into the distant future, the concept of recovering energy from waste is becoming particularly attractive. In addition, solid waste can be considered a low sulfur fuel.

Solid wastes can be burned to produce steam that can be used for heating purposes, or to generate electricity. Also, solid waste can be converted to pyrolysis gas or oil that would be storable and/or transportable. Burning of solid wastes to produce steam is the best short-term solution, but the best long-term approach could be the production of pyrolysis gas.

WASTE HEAT RECOVERY

A good summary of the situation has been extracted from a report by Richard B. Engdahl of Battelle Memorial Institute under a contract with the EPA, and is given below.

A few combustible industrial wastes generated in manufacturing processes have normally been used as fuel by the organizations that generate them. Useful heat is required for generation of process steam or electric power. Usually, this has been done in conventionally designed boilers and furnaces, with perhaps specialized equipment installed only for conveying and feeding the combustible material into the combustion chamber.

Similarly, the recovery of useful heat from the incineration of mixed solid municipal wastes and sewage sludge has been practiced extensively in Europe, but not in the United States because of the generally unfavorable fuel economics for waste heat recovery. If in addition to whatever fuel saving is possible, however, some value were placed on ease of handling of a refuse disposal problem and on environmental quality (air purity and land beautification), heat recovery from the combustion of municipal waste could be made more attractive. The outcome of each analysis will depend on specific local conditions and standards of environmental quality. In many locations waste heat recovery may not seem justified for many years.

Heat Available: The long-term trend has been for the cellulose content of municipal solid waste to increase with decreasing moisture; hence, the heat value from combustion of the waste has been rising. In the United States, it was estimated to be about 5,000 Btu/lb. This trend will probably have some effect on increasing the feasibility of waste heat recovery.

Methods of Heat Recovery: The simplest method of waste heat recovery from municipal incineration has been the incorporation of a waste heat boiler immediately following the incinerator for the regeneration of hot water or steam. One disadvantage of such a system has been that the refractory-lined furnace would be unable to withstand the temperature generated by stoichiometric combustion of the waste; hence, the chamber must be cooled by dilution of the combustion gases with excess air.

The logical alternative to this less efficient system is the construction of an integral boiler, the furnace lining being formed by steel tubes to provide a water-cooled incinerator chamber. This has been demonstrated to be highly successful in many installations in Europe. One difficulty has been with boiler-tube corrosion due to sulfates or chlorides on the fire side of the tubes. This attack has appeared to be a function of steam temperature; that is, if steam temperatures can be held below 1000° F., little or no tube attack should occur.

An alternative method of heat recovery is to have an air-heater type of heat exchanger following the incinerator chamber. Again, this might be very inefficient because excess air would be needed to cool the flame to protect the refractory walls of the incinerator, and the temperatures of the gas-to-air heat exchanger metal would have to be held to moderate levels to give reasonable heat exchanger life. All of these methods must be designed to cope with extremely dusty gas probably involving some adhesion of the dust to the tubes.

Applications for Recovered Heat: If the waste heat is recovered as steam, it can be utilized for turbine drives for the auxiliary equipment on the plant, such as blowers and pumps. Alternatively, the turbine can drive an electric generator for the generation of electric power for internal use in the plant or for sale. If the plant is located adjacent to a populated area, the steam can be used for district heating or for industrial processing. A plant has been installed for combination of seawater distillation with incineration. Instead of steam, the heat can also be recovered in the form of hot water, which can be used for heating the premises or for sale in the neighboring area. This has been done extensively in Denmark.

Performance of Heat Recovery Systems: The amount of steam generated per pound of refuse burned depends on many factors. The most efficient generation has been in water-tube-wall boilers operating with low excess air without interruption for 24 hours a day. In general, to achieve satisfactory heat generation, it has been necessary to provide auxiliary fuel to maintain constant generation because of the varying moisture content of the refuse and the varying supply of refuse. In general, the reliability of a waterwalled waste heat boiler will be highest of all, unless tube temperatures are carried so high that corrosion is a problem.

Economic Incentive for Heat Recovery: The value of the heat recovered will be determined by the cost of using competitive fossil fuels and upon the load factor that needs to be maintained for the particular use intended. A major economic advantage to the recovery of heat, particularly where environmental quality is considered important, has been the volume effect of the extraction of heat from exhaust gases in the furnace and the use of less excess air because of the completely water-cooled furnace. Both of these factors have greatly reduced the volume of dusty gases to be cleaned, and hence have reduced first costs, operating costs, and maintenance costs of suitable dust collecting equipment. (End of Engdahl report.)

WASTE PROCESSING

Most schemes for processing waste, prior to firing, provide for the removal of magnetic materials. However, if additional noncombustible material is removed, the quantity of bottom ash will be reduced, the Btu value of the fuel increased, and there should be less abrasion and/or erosion. A process developed by Combustion Equipment Associates provides for multistage classification for removal of additional noncombustibles, two-stage shredding, and drying. This process is schematically illustrated on the following page.

Current Activity: Numerous studies are being conducted in the United States as to the suitability of recovering heat from solid waste. There are as yet no power generating incinerators in the United States, but the cities of Chicago and St. Louis each have a steam generating incinerator, and another in Nashville is planned to go on-stream in early 1974.

PYROLYSIS PRODUCTS

From an energy yield standpoint, gas pyrolysis offers an advantage over oil pyrolysis. However, the gas pyrolysis plant is best adapted to location near a large energy user in order to minimize pipeline and storage costs. Pyrolytic oil, on the other hand, offers the advantage of more economic storage and distribution by truck, rail, or barge. This allows greater freedom in locating the plants near the centers of waste generation. Pyrolytic gas can be upgraded to pipeline quality but requires an increase of approximately 50% in capital costs. If this is done, more freedom in location of pyrolytic gas facilities is provided. The only limitation then becomes the need to connect to an existing natural gas pipeline for distribution of the final product. Pyrolysis processes permit cleaning of the oil or fuel gas being produced, thus reducing the air pollution and boiler corrosion problems that limit solid refuse fuel to use as a supplementary fuel.

A demonstration oil pyrolysis plant is now being designed by the Garrett Research and Development Company for installation in San Diego County. This

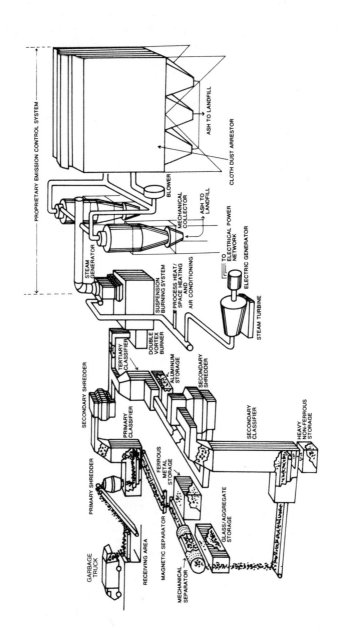

Introduction

200 tpd plant is jointly funded by an EPA grant, San Diego County, San Diego Gas & Electric, and the Garrett Company. The design is expected to be completed by the end of 1973, with construction during 1974. Start-up operations are expected to require the first four months of 1975, followed by routine operations.

Union Carbide is now building a 200 tpd demonstration plant in South Charleston, West Virginia, to prove out the key components of their gas pyrolysis process. This will be funded by Union Carbide in an existing Union Carbide plant. The demonstration will be located in its own building, and will utilize blown off waste oxygen from other processes in the plant. The demonstration plant will utilize an 11 foot diameter shaft furnace and is expected to start up by early 1974.

Garrett and Union Carbide are not the only developers of pyrolysis processes. Their schedules, however, were felt to be roughly indicative of the fact that availability dates for full scale pyrolysis processes with pilot plant experience are 1977 to 1978.

Refuse As a Low Sulfur Fuel

The information given below is based on a report prepared for the EPA in 1971 by Envirogenics Company, Foster Wheeler Corporation and Cottrell Environmental Systems, Inc. entitled *Systems Evaluation of Refuse as a Low Sulfur Fuel.* Costs are based on 1969 figures.

REFUSE AS A FUEL

Quantity-Quality Considerations

It is estimated that the U.S. population will have increased by almost 60% by the end of the century. Because of the changing living styles and other factors, the per capita production of mixed municipal refuse will also increase. On the most probable basis, the national "resource" of refuse will have increased by a factor of 2.8, or from about 160 to 450 million tons per year, between the years 1965 and 2000. A factor as high as 4.0 is also possible.

With the introduction of more garbage disposal units, the substitution of more plastic for glass and metal goods, and the acceptance of more new items of short-lived (one-time-use) merchandise, the composition of urban refuse can also be expected to undergo change. The following tabulated predictions are considered to be reasonable.

Projected Compositional Changes in U.S. Urban Refuse

Composition, weight percent

Year	Garbage	Plastics	Glass	Metals	Garden	Paper	Residual
1970	20	2	12	10	12	38	6
2000	6	13	7	5	12	55	6

The composition of municipal refuse at a given location varies fairly widely in both the short and long (seasonal) scale. Physical analyses of solid waste over a period of a week or so in the same area show wide variations and seasonal effects on such constituents as garden wastes account for most of the variation through the year. The most striking thing about the geographical variation in refuse composition is that it is so small.

Aside from the geographical effects on amounts and peak-appearance of garden wastes, municipal refuse is much the same the country over. This may merely be caused by the homogeneity of the culture in the U.S. From the point of view of refuse as a fuel, its physical analysis is unimportant except insofar as it determines the heat of combustion and, to a degree, corrosion potential. Because the seasonal variations in waste load result largely from low Btu, high moisture yard and garden wastes, the approximately ±8% variation in average waste load introduces only about a ±6% variation in total heat available from refuse over a year in a typical case.

The heat of combustion of refuse has grown steadily over the past several decades, but will probably not grow as fast during the next three. Based on the compositional changes shown above, an approximate 30% growth, from about 5,000 Btu/lb. in 1970 to about 6,500 Btu/lb. in 2000 AD, can be expected.

Reduction in Pollutant Emissions

At the present time, if all the urban refuse generated in the U.S. were to displace an equivalent amount of coal energy, approximately 25% of the coal now fired in power plants could be saved. By the year 2000 this displacement will probably have increased to almost 40%. This projected increase will be due not only to the enhancement in fuel value that refuse is undergoing, but to the probability that, on a per capita basis, refuse generation will increase at a faster rate than the demand for electricity that can be satisfied by fossil fuel energy. This prediction may prove to be in serious error if the problems now encountered in the construction of nuclear power plants persist.

It is well-known that the sulfur content of refuse is low in comparison with that of coals and residual oils. Thorough review of the literature shows a consistent average sulfur content of 0.1 to 0.2% in U.S. refuse. This contrasts with a range of 2.5 to 3.5% for those bituminous coals most commonly being fired in power plants. Correcting for the differences in heats of combustion, sulfur input to the boiler could be reduced by a factor of from 5 to 15. It is also well documented, however, that substantially all (95 to 100%) the sulfur in coal or oil fired to a boiler will appear in the flue gas as the oxides.

Data available for refuse incinerators indicate that only somewhere between 25 to 50% of the input sulfur is released as SO_2. This is doubtless due to the fact that a significant portion of the sulfur present in trash is in the inorganic salt or fixed form. Thus if all the urban refuse now available in the U.S. were to displace 2% S coal, over 2 1/2 million tons of SO_2 would be eliminated from the atmosphere annually. This is based on the premise that no SO_2 control would be practiced, which is essentially the case now, and that the percent of input sulfur released as SO_2 is 95 and 50%, respectively, for coal and refuse.

Evaluation of data on particulate release rates for power plants and refuse incinerators indicates that about the same amount (2 1/2 million tons per year) of particulates as SO_2 could be eliminated. This, however, is also based on a no control situation which is obviously unrealistic. Only about 120,000 tons per year would be eliminated if all the power plants involved were equipped with

APC equipment having an average efficiency of 95%. The reason for the higher particulate release rates for coal is explicit from the modes of firing.

Refuse is usually burned, without prior size reduction, on grates, while coal is fired in pulverized form in suspension. If shredded and fired in suspension, refuse would doubtless generate more fly ash, but certainly not to the extent of 80% or more of its inert content, which is typical for coal. The ash particles formed would still be much larger from shredded refuse than from pulverized coal, and would contain large amounts of dense materials (e.g., glass and metal) that would be less susceptible to elutriation than coal ash.

Because total consumption of the national refuse inventory in power boilers is unlikely, the more realistic model is the LMA. Using the St. Louis area as an example, it was found that if all the available refuse there were fired in power boilers to replace part of the coal, there would result a SO_2 reduction of about 31,000 tons per year. Based on the SO_2 burden from all sources, this would represent a reduction of about 7%. The reduction would only amount to 0.5%, however, if the power plants were equipped with SO_2 scrubbers of 95% efficiency. In the case of overall particulate burden, a reduction of 21% would be experienced if no APC equipment were being used on any power boiler. This reduction would be only 1%, however, if particulates from all power plants were being removed by air cleaners with a 95% efficiency.

Coal displacement by combination firing with refuse will in itself obviously not completely solve the SO_2 problem and clearly the development of APC systems for removing this pollutant will still be necessary. An important aspect of the refuse utilization concept is the mode in which it would be fired. The argument could be raised that the firing of refuse with reduced amounts of coal in a single furnace chamber would actually provide little, if any, benefit. Scrubbers would still have to be sized on the basis of flue gas volume and this would increase as the refuse proportion was increased. Annualization of this added cost might then actually exceed any operational cost reduction that would be realized by a diminished rate of sorbent exhaustion.

As will be discussed later, however, the preferred configuration of the combined-fired power boiler involves an almost complete isolation of the refuse and fossil-fuel furnace components. Thus, in terms of SO_2 removal, the volumetric cleaning requirement would be reduced in a linear manner with respect to the amount of coal displaced by refuse.

Thermal Utilization

While the obvious application of a combined fired boiler would be the production of electricity, other energy utilization schemes should not be overlooked. In fact, refuse energy need not even be exclusively considered in the context of the Rankine Cycle. This possibility is now being explored with the CPU-400 System, which consists of a fluidized-bed refuse furnace used as a gas generator to drive a gas turbine and coupled power generator. Because of its in-development status, however, this system, as well as others, was not considered to be a candidate ready for the analysis.

A number of alternate uses for steam generated by refuse (or any fuel, for that matter) have been reviewed. None was regarded as offering a strong challenge to the power plant concept, yet most would probably be acceptable if a favorable application situation existed. For example, district heating would be an attractive means of dispensing refuse energy, although during summer months an alternate form of heat extraction (e.g., through absorption refrigeration) would be necessary. Evaporative water desalination units would also be excellent devices in which to convert refuse to credits. However, only a limited number of such installations would likely be built within reasonable reach of a municipal refuse-collection system.

POWER PLANT DESIGNS

A survey of the state of the art of domestic and foreign power plant designs was conducted with the aim of identifying those most amenable to combined firing of refuse and fossil fuel.

Pulverized coal-fired utility steam generators in service include horizontally and tangentially fired furnaces, both wet and dry bottom, as well as cyclone furnace units; however, units being sold now are virtually restricted to dry bottom, horizontally or tangentially fired systems. In this country, natural circulation boilers are now the dominant design form in the size range appropriate for refuse-firing (500 MW or less because of refuse logistics). In Europe, however, the once-through system still appears to be the favorite.

From the point of view of combined firing of municipal refuse in utility-type steam generators, no experience is, as yet, available in North America. This practice has grown in Europe, almost entirely in the last decade, and is most advanced in Germany. European practice involves reliance on various refuse burning grates, with pit and crane refuse handling, but with a variety of steam generating arrangements.

All presently operating, large North American steam-generating, refuse disposal plants use conventional, agitating grates and generate steam at such moderate pressures and temperatures as to be unsuitable for turboelectric applications. All except the incinerator at the Naval Base at Norfolk, Virginia, fire refuse only; the Norfolk unit fires a combination of oil and the solid wastes collected at the base. It was, however, the first water-walled refuse incinerator built in North America. Water-walled, steam-generating incinerators are now under way in Hamilton, Ontario; Braintree, Massachusetts; Harrisburg, Pennsylvania; and Chicago, Illinois. A recently completed plant in Montreal is in its initial shakedown operation. Again, all of these systems are low or moderate pressure units.

Current domestic technology in bark, bagasse and wood chip fired heat recovery furnaces is pointing the way to advances over the present state of the art of refuse combustion, which is almost universally based on the use of the reciprocating or turning grate. The installation now under construction at East Hamilton, Ontario, will utilize the next logical advance in refuse combustion, namely the spreader-stoker equipped with a burnout grate. Planned for test in the near future in St. Louis is the suspension firing of shredded refuse in a grateless furnace, actually an old, coal fired utility boiler modified for such service. A similar system is under construction in Rochester, New York, although it will have a burnout grate, as does its bark-burning prototype at Muskegon, Michigan.

Variations in steam flow of ±30 to 40% have been observed in grate-fed steam generating incinerators both here and abroad. These fluctuations can have periods of from 15 to 25 minutes, with maximum Δp-rates of from 2 to 8% per minute. These wide fluctuations can be greatly reduced by suspension or spreader-stoker firing of ground refuse. This would result from the combined effects of size reduction, mixing, better contact with combustion air, and the reduction in furnace load of unburned fuel.

Another engineering approach, particularly desirable if the agitating grate is to be used, would be to employ separate refuse and coal-fired furnaces. The latter, which are far more susceptible to control, could thus be used to develop the final steam conditions required for the system, and thus even out fluctuations introduced by the refuse-fired portion of the boiler. A secondary benefit would be that the refuse furnace could then be operated at lower steam temperatures. This would serve to reduce the corrosion effects that sometimes occur on tube surfaces in refuse-fired boilers when steam temperatures are allowed to exceed 800°F.

The conservative viewpoint in the U.S. has been that corrosion resulting from refuse combustion would not permit its use in modern, high pressure, high temperature, steam generators. Reports of severe corrosion in the several modern combined firing steam generators in Europe circulated soon after their start-up, and are still repeated. Discussion by project personnel with the plant operators and others confirms an opposing view that tube wastage, while expectedly greater than with fossil fuels, is a nuisance rather than a severe problem. Tube-wall corrosion rates were initially high but soon levelled off to tolerable rates, apparently after the tubes became coated with protective deposits.

Based on the considerations discussed above, the boiler designs developed on the program were limited as to refuse input, generally at 60% of the energy. Steam conditions were also approached on a conservative basis, particularly as the refuse input approached the specified limit. The steam generators and other components sensitive to flue-gas flow rates were sized for 50% excess air for refuse-firing and 18% excess air for the coal. These criteria were established on the basis of German practice. Design exit-flue-gas temperature was set at 450°F., which is considerably higher than current practice in utility power plants. This temperature selection was based on performance constraints imposed by the use of an electrostatic precipitator and the resistivity characteristics of refuse fly ash.

Based on the holding limits of commercially available grates, the maximum unit refuse capacity adopted was 1,000 tpd for furnaces equipped with agitating grates and 2,000 tpd for those with sloping, fall out retainer grates or none at all. An upper limit (for analytical purposes) of refuse input per plant was arbitrarily set at 8,000 tpd, although it was recognized that a system of such great size would only be feasible if unique refuse in-haul techniques were utilized.

The capacity range for individual units was derived along similar lines and the limits set at 85 and 500 MW, the latter being highly unlikely of attainment unless a very low refuse proportion were employed. Because nameplate rating would not exceed 500 MW, it was then logical to utilize natural circulation steam circuitries. Systems having capacities of less than 85 MW, which would perhaps be appropriate for smaller communities, were not considered because refuse disposal costs were found to be unacceptably high at the low refuse rates entailed.

Another aspect of the design approach included provision for both SO_2 control (limestone wet-scrubbing with flue gas reheat) and dust control. The latter would be handled by an electrostatic precipitator for the flue gas from the refuse-fired furnace and by the limestone wet scrubber in the coal furnace flue. If the flue gases were blended prior to the air cleaner, an appropriately sized wet scrubber would alone be used to remove SO_2 and particulates for the entire system. Control of HCl and certain other gaseous emissions identified with refuse incineration did not fall within the scope of the study. Such control may well develop as a requirement of the future, as the composition of refuse changes.

Refuse handling procedures commonly practiced in U.S. incinerators and even in European steam-generating plants are based on a flow arrangement wherein bridge cranes move the feed from furnace-house storage pits to the furnace charging chutes. Alternative, more cost-effective designs are now being explored. These are for systems for which all input refuse must be shredded. Live-bottom receiving and storage structures have been specified and the movement of the shredded feed material is by means of conveyors or pneumatic pipelines. In the study, it was found desirable to specify similar handling equipment whether the input were to be shredded or not.

New Plants

The capital and operating expense of grinding refuse had a dominating influence on the disposal costs for the various designs analyzed. Those candidates employing agitating grates wherein only oversized input items would have to be reduced were generally found to be the more cost-effective. Within this class, a configuration designated as Case 3, clearly emerged as an optimum design for the unit size range (200 to 400 MW) best suited to utility and waste management operations in large metropolitan areas.

Despite indicated higher disposal costs, a candidate was also selected from the class of designs specifying suspension firing. This was done because, lacking any data from precedent refuse-firings, conservative performance characteristics had to be assumed for these systems, and because refuse-grinding costs needed further verification for similar reasons. The system selected from this class, designated as Case 10, was seen in the cost modeling to be optimum for that type of design. Summary descriptions of the two class-optimum candidates follow.

Steam Generator with Refuse-Fired Economizer (Case 3): The Case 3 steam generator design resembles, in basic layout, a widely regarded combined-fired boiler now in operation in Germany (Munich South, Unit No. 6). The two differ significantly, however, in several major features, as will be brought out later. At the sizing (400 MW) shown by the cost model to be optimal, the Case 3 system comprises a pulverized coal fueled boiler equipped with very little economizer surface but otherwise of conventional design, and three refuse-fired furnaces that would serve as the actual economizers.

These feedwater heaters would burn about 930 tpd of refuse on reciprocating grates and deliver feedwater slightly below saturation (657°F. and 2,600 psig) to the economizer (actually a mixer) of the steam generator. The latter, firing about 3,000 tpd of pulverized coal, would produce about 2.8×10^6 lb./hr. of steam at 1000°F. and 2,400 psig for a single turbine-generator set equipped for reheat operation. The net plant heat rate would be about 10,100 Btu/kwh.

The flue gas from the economizers would be intrinsically isolated from that of the steam generator. Thus the flue gas exiting from all three economizers would be combined, passed through an electrostatic precipitator, without concern for SO_2 removal, and on to a common plant stack. In the coal-fired boiler, the flue gas would be passed through a limestone wet-scrubber to remove SO_2 and particulates and thence to the common stack.

Based on the cost model criteria discussed earlier and a net plant refuse rate of 2,790 tpd (25% of the energy input), the total net disposal cost would be $1.61/ton. This figure includes, however, a transportation cost of $1.65, showing that the plant itself would operate at a profit of $0.04/ton. Thus refuse disposal using this particular design would not only be competitive with but considerably more cost-effective than landfill.

Steam Generator with Refuse-Fired Arch Furnace (Case 10): The system selected from among the suspension-fired designs as being the most cost-effective would incorporate a refuse-fired arch furnace. In this unit, ground refuse would tend to burn in suspension because of the mode of injection and reaction to the lifting effect of directional air flow. Water-cooled grate bars would be installed on the sloping walls of the hopper to provide some retention of fallout. The need for this feature is highly uncertain and its omission would greatly reduce construction costs.

In the total plant system, 1,960 tpd of refuse (60% of total energy) would be fired in conjunction with 496 tpd of coal. A single arch furnace would fire all of the refuse in combination with 182 tpd of pulverized coal, the latter to insure even combustion. This unit would produce 0.8×10^6 lb./hr. of steam at 750°F. and 1,940 psig. The steam would be fed to a coal-fired (164 tpd) superheater of special design, where it would be brought to final turbine inlet conditions (1000°F. and 1,800 psig). A single turbine would be needed to satisfy the design goal of 100 MW.

Because of tube-surface distribution constraints, reheat for the turbine would be accomplished in a second coal-fired (150 tpd) furnace rather than in the superheater. On the fire side, the two coal-fired systems would be almost identical. A number of design differences would be necessary for the steam side, however. Controls for such a system would necessarily be more complex than for conventional steam generators.

The disposal costs for the suspension-fired systems were found to be higher than for grate-fired configurations. In the present case, the estimate came to $6.17/ton, including transportation costs. The accuracy of this value is much less certain than for the more straighforward Case 3 situation. Not only might the performance specifications, but grinding costs as well, prove overly conservative. The data available for developing the latter were quite limited and tended to be inconsistent. The grinding costs used for this design (2 inch top size) were about $2.50/ton, including annualization of capital costs, and operating and maintenance costs.

Retrofit Plants

The plans developed for retrofit opportunities were based on examples of existing, utility-class boilers that had gone into service in the earlier part of the past two decades. Within this class of older boilers, only the more typical designs and steam conditions were considered. Five retrofit designs were prepared for unit nameplate ratings ranging from 44 to 300 MW. In each case, the design was so developed that the original unit capacity and steam conditions were not changed.

Modification No. 1: The selected 60 MW unit (no reheat capability) firing pulverized coal from horizontally-aligned guns would be retrofitted with a reciprocating grate. Extensive changes in the steam circuitry in the lower portion of the furnace were indicated, as was the addition of new air supply equipment. The modified unit would derive 43% of its energy from refuse (951 tpd) and produce, as before, a steam output of 0.6×10^6 lb./hr. at 900 psig and 900°F.

Modification No. 2: In this design, a separate, refuse-firing furnace, equipped with reciprocating grate, would be added onto an existing 150 MW unit and the two furnaces coupled to share the existing convection and downstream flue structure. It is assumed, of course, that sufficient space is available at the site to permit such an expansion. This retrofit system would generate 1.1×10^6 lb./hr. of steam at 2,035 psig and 1053°F., using a reheat cycle. About 24% of the energy would be derived from refuse, based on a rate of 1,000 tpd. Because of the add-on boiler effect, the cost of this particular modification proved to be much higher than for any other retrofit system considered. The resulting unit would be similar in function to boilers now in operation in Munich and Stuttgart.

Modification No. 3: This retrofit design would involve a restructuring operation similar to that specified for Modification No. 1. The grate installed, however, would be of the travelling type. Air-swept spouts would be installed above the grate for introducing shredded (4 inch top size) refuse into the furnace; thus, a spreader-stoker configuration would be realized. This 44 MW unit would generate 0.4×10^6 lb./hr. of steam at 1,370 psig and 905°F. About 50% of the energy would be derived from refuse (846 tpd).

Modification No. 4: This design was based on the use of the same existing plant just considered for Modification No. 3. The boiler would, however, be equipped with a reciprocating grate and operate in much the same manner as Modification No. 1. The basic difference between the latter and the present system would be in the styles of construction of the existing boilers and therefore the modification techniques necessary to achieve conversion. The existing plant considered for Modification No. 1 was a much older plant than the present one, such that considerably different approaches would be necessary to produce what would essentially be the same result. The design calls for firing of 679 tpd of refuse, which would represent 42% of the heat input.

Modification No. 5: This design involves the modification of one of the units at a midwestern power plant where a similar retrofit is already being pursued under EPA sponsorship. In the present case, suspension firing would also be practiced using refuse nozzles that would be wall-mounted above existing, horizontally-aligned coal guns. A notable difference would be the conversion of the hopper walls to provide a water-cooled grate surface. Being steeply sloped, this surface would only have a delaying effect on, but thus promote more complete burn-out of, fallout tumbling into the ash pit. This conservative design feature may prove unnecessary, in which case a major modification cost element could be eliminated.

The hopper-wall treatment of the present retrofit system, that of Case 6, which is identical in principle, and that of the arch furnace (new construction, Case 10) are very similar and provide no horizontal retaining surface. Thus the completeness of particle combustion while in suspension is an important consideration in determining the refuse fraction to be used. Burn-out in an arch furnace is expected to be more complete because of furnace air distribution. For this reason, a lower refuse fraction has been stipulated for the present system. Based on a conservative fractional heat input from refuse of 10%, this 300 MW unit would consume 635 tpd of that fuel. Steam production would be about 2.3 x 10^6 lb./hr. at 2,200 psig and 1010°F. using a reheat cycle.

Disposal Costs of Retrofit Systems: A different approach than that used for the new construction units was necessarily employed in deriving disposal costs for the retrofit designs. Annualization factors for the existing plant prior to retrofit were of course excluded, as were credits for electricity generated. Many of the cost derivations employed in the cost model were applicable, however. The results are tabulated below.

Refuse Disposal Costs for Retrofit Systems

	Mod. 1	Mod. 2	Mod. 3	Mod. 4	Mod. 5
Disposal Cost, $/ton	0.65	2.87	2.14	1.07	3.69* or 4.44

*Lower cost results if hopper walls are not modified.

The above data again show that suspension firing (Modification No. 3 and No. 5) leads to apparent higher disposal costs. Of the grate-stoked systems, Modification No. 2 proved to be an exception to this effect because of its very high reconstruction cost. An important parameter that is included in the disposal costs shown above is plant factor, which was 80%. This is the same value that was applied to the new construction designs and may prove overly optimistic considering the age of the retrofitted boilers. Disposal costs would increase about 25% for each 10% (relative) drop in plant factor until a plant factor of about 50% is reached.

ENERGY UTILIZATION

Turboelectric Generation

Rankine Cycle Systems: All, or a portion, of the thermodynamically available energy content of steam generated from external combustion of refuse can be converted to electricity through expansion of the high temperature and pressure steam through any of a variety of turbine systems. The refuse can serve as the sole heat source or be used in combination with fossil fuels.

Gas Turbine Systems: Rather than transfer of heat to a working fluid as is accomplished in the Rankine cycle, it is possible to directly use hot combustion gases to impart rotary motion to a gas turbine. Distillate oil-fired gas turbines have become common in the electrical utility industry, particularly for relatively small units for the rapid supplying of peak energy demands.

A number of refuse-fired gas turbine systems have been suggested. The one receiving the most attention is that being developed by the Combustion Power Co., the so-called CPU-400. This unit, intended to produce 15 MW of power from the combustion of 400 tpd of solid waste, also yields 250 lb./sec. of gas at 950°F. that can be used for process heat, sewage sludge drying, desalination, etc. The design incorporates shredding, air classification for the removal of inerts and heavy objects, and partial drying of the refuse using a small amount of the waste heat.

Combustion in the CPU-400 system occurs within a 100 psia reactor, where inert sand particles are fluidized by means of 584°F. air from the compressor. Fly ash and fines from the fluidized bed are removed prior to introduction of the 1650°F. gas into the turbine by inertial separators and an electrostatic precipitator. Ash that is heavier or larger than the bed sand will be periodically removed from the reactor bottom through a rotary air lock. Each of the critical hardware elements has been tested with subscale equipment and further developmental work with larger units is in progress.

Full economic evaluation of this concept cannot be completed until additional engineering tests indicate the degree of complexity of the combustion and gas cleaning equipment, turbine life, etc. Preliminary costs have been estimated by the Combustion Power Co., but direct comparison with values derived on this program should be avoided. The capital cost of $4 million corresponds to $10,000/tpd incinerator capacity or about $375/kw. installed capacity as electric plants are evaluated. This compares with $120 to $140/kw. for present stations in the same size range. Under private company ownership, it is estimated refuse disposal costs would total $5.99/ton, without credit for by-product sales. With credit for electric power (82% load factor and 4.5 mills/kwh), disposal costs are estimated at $2.67/ton. Municipal ownership is claimed to lower these values to $4.27 and $0.95/ton. Utilization of the turbine waste heat would further reduce these costs.

Other Applications

Heat from the combustion of refuse may be used in numerous ways to benefit the surrounding community. In addition to the obvious use for electrical power generation, there are several other potential end uses that deserve consideration. Among these are industrial and district heating (and cooling), desalination, and various miscellaneous applications. These uses utilize steam of a relatively lower grade than that required for power generation, and in some cases could use the steam discharged from another process. The secondary heat sources that may be considered in this respect are: (1) the steam bled from one or more stages of an extraction turbine; (2) the discharge from a back pressure turbine; (3) output from a low pressure boiler; and (4) steam discharged from some higher level process. The last option may permit the cascading of two or more uses in series.

Industrial and District Heating: Heat from refuse could be used by many different industries, ranging from chemical processing plants to food packing to commercial laundries. The form of energy required varies over such a broad range that each particular industry would have to be evaluated on an individual basis. This is not to imply that industrial heat should be disregarded as a potential use for refuse energy, but that the type and density of the surrounding industries may well dictate the steam conditions and steam generating equipment required. District heating has become of increasing importance for space and hot water systems. Tremba, in his recent review of all known district heating systems, indicates the output magnitude of present installations.

During the years 1965 to 1967 some 2.9×10^{15} Btu were distributed, 88.5% within the Soviet Union and 9.1% within the western world. In combined heat and power plants, approximately 1.1×10^{14} watt-hours of electricity were produced. Of the 226 heating installations surveyed in the western world, 21 stated they used refuse as a portion of the fuel for the boilers. The siting of power facilities within cities has led to low distribution costs of both heat and electricity in Russia, with consequent extensive employment of district heating. In Moscow, for example, 22,000 buildings and 360 industrial plants are connected to a heating system containing 750 miles of pipe.

In Denmark, 30% of all dwellings are now heated from central boiler plants, and this is expected to increase to 40 to 45% by 1975. About 450 plants have been built, and operation has been successful at dwelling densities as low as 4/acre. Normally, pipe lengths from the boiler to the farthest consumer are less than 2.5 miles. It is estimated that an annual 38 lb. SO_2/dwelling connected to the heating system is no longer discharged at the previous low levels of home chimneys. Selling price to the small user in Denmark is in the range of $0.80 to $1.60/$10^6$ Btu which accounts for the popularity of these heating systems (allowance should be made for the elimination of capital and maintenance costs of individual home furnaces and water heaters in comparing such heating costs).

Mørch estimates that in Denmark a city's refuse would cover 6 to 8% of the heat consumption over the year, and that if the usual wider refuse area is considered, this percentage would be significantly greater. A test facility has been built for producing gas from refuse (7 cu. ft./lb.), which, after carbureting with 3.4% butane, has a heating value of 510 Btu/cu. ft. Evaluation is now being made for use of the plant waste heat plus the gas manufactured in a district heating plant, as well as supplying a portion of the gas to city mains.

Kimura gives emphasis to the need for an industrialized, high population density nation such as Japan to develop a plan for optimum utilization of all energy factors. Within his recommended "circulation network of city energy," the burning of refuse plays an important role in a number of heat generating systems. In his review of district heating systems in the Mannheim area, Winkens described that city's Nord plant in which 80 tons/hour of refuse is fired in two boilers in addition to three oil-fired boilers. Steam is supplied to transmission grids at both 20 and 7 atm. pressure, with some 250,000 tons being supplied in 1969 from the incineration of 130,000 tons of refuse. Expansion of the waste burning facility is planned in the near future.

Britain's first combined refuse incinerator and district heating plant is being constructed in Nottingham. The output will increase through planned expansions to 440 million Btu/hr. by 1980. At this latter time, some 170,000 tons of refuse per year would be burned, providing up to two-thirds of the total heat output (remainder from coal). Steam from separately-fired coal and refuse furnaces will be used for electrical generation and the supplying of hot water at 285°F. and 85 psi to distribution mains. Reduction of about one-third in the cost of heating and hot water are anticipated.

The Northwest incinerator plant in Chicago, now under construction, will utilize a portion of its steam for district heating. An average production of 3 lbs. steam per lb. refuse is anticipated and the facility is designed to handle 560,000 tons per year of refuse. Most feedwater pumps and fans at this plant will be steam turbine driven once start-up is achieved.

The most advantageous applications of absorption refrigeration are in cases where suitable amounts of low-grade heat are available, such as can be the case with a refuse-fired boiler. Where reciprocating compression machines are limited by volumetric displacement, the absorption machines can maintain capacity at lower back pressures by increasing the flow of heat to the generator. Absorption systems generally require no special building considerations. With the application of suitable controls, the absorption system will operate with very little attention. Absorption refrigeration systems have been built in capacities of 2,000 tons (24 $\times 10^6$ Btu/hr). The absorption system does not respond well to rapid changes in load, but should operate nicely in a district cooling application where the load is averaged by the distribution system. Conventional compressor driven refrigeration or heat pump systems could also use steam turbines as the driving device in large district cooling or heating systems.

No detailed analysis has yet been made of the size range over which economical operation of district heating from refuse combustion might be realized. Rather typical of the qualitative analyses that have been made is that of Beningson, where wide application of incineration energy is encouraged, but not justified on an engineering or economic basis. A generalized economic optimization of a dual-purpose (heat and electricity), multifuel (refuse and fossil), steam generator is affected by a large number of variables, chief among which are:

(1) Regional electrical demand and anticipated growth rate
(2) District (local) heat demand and anticipated growth rate
(3) Distribution of population density (cost of transmission system)
(4) Temporal effects on output demand for both services
(5) Annual outdoor temperature statistics
(6) Fossil fuel costs (heating basis)
(7) Refuse haul and processing costs
(8) Refuse heating value
(9) Factor costs (capital charges and labor).

When considering the cost of the transmission system, it should be noted that it is essential that the heat consumers be near the generating facility, while this is not the case for the users of the electrical energy. However, a privately owned dual-purpose plant can still compete with grid power in spite of higher electrical generating costs when their distribution costs are small.

The above determinants will indicate the basic steam cycle that will lead to lowest annual costs for the total system (refuse disposal, heat supply, and electrical generation), which in turn would permit firm design of the furnace-boiler-turbine-condenser components. At the 1st International District Heating Convention held in London in 1970, a number of suggestions were made for optimization techniques for dual-purpose plants using fossil fuel only.

Adaptations of these could be made for the more complex case with waste fuel for a specific set of assumptions.

Desalination and Miscellaneous Applications: Waste heat can serve as an excellent evaporation source for a desalination facility. Plant optimization depends upon factor costs, but a typical yield of fresh water from brackish or seawater is about 10 lbs./lb. steam. In areas where natural water shortages exist, the use of the energy from refuse combustion should definitely be considered, particularly after a portion of the energy has been extracted in a turbine system.

Steam derived from fossil fuels has been used for snow and ice melting on sidewalks, roads, and airport runways; properly designed systems using refuse as a fuel should offer significant cost advantages. Other potential applications include the use of refuse-derived steam as the motive power for driving pumps, fans, and ejectors. Where linear actuation is required, steam has been satisfactorily used for purposes ranging from clamping and punching operations to the catapult launching of aircraft. Acoustic generators of many types are used in industrial applications. Considerable interest has developed in jet cutting of many materials and here steam could supply the required energy for many such operations.

St. Louis Project

The particular process described in this chapter is the recovery of thermal energy by burning shredded residential solid waste as supplementary fuel in boiler furnaces and is based on a report by Horner & Shifrin, Inc. This process was initially studied in 1968 by the City of St. Louis with partial funding from the Federal solid waste management program. The results of that study were encouraging; the required equipment is commercially available, and a power company serving the metropolitan area, the Union Electric Company, concurred with the high probability of success of the process by offering to the City its full cooperation. This included providing two test boilers and offering to contribute a substantial share of the equipment, construction, and evaluation costs if a full-scale demonstration program could be arranged.

RECYCLING RESIDENTIAL SOLID WASTE AS SUPPLEMENTARY FUEL FOR POWER PLANTS

In April 1972, for the first time in the United States, and perhaps for the first time in the world, an investor-owned utility began to burn municipal solid waste as fuel for the direct production of electric power. This unique cooperative venture, between the City of St. Louis, Missouri, and the Union Electric Company, with financial support from the Environmental Protection Agency, is intended to demonstrate the benefits that will accrue to a metropolitan area using this resource recovery technique for solid waste management. The potential benefits include an environmentally acceptable means of solid waste disposal, conservation of natural resources, more effective control of land use, and economic advantages to both the utility and the public. The objective is to achieve all of these benefits by applying existing technology, with equipment which already is commercially available.

Perhaps the most striking aspect of the process is its simplicity. Domestic solid waste, collected from residential areas of the City of St. Louis, is ground up in a large hammermill, magnetic metals are removed, and the remainder of the material is fired pneumatically to one or both of two existing boilers in the Union Electric Company system. Glass, ceramics, and other nonmagnetic inert materials will not be removed from the refuse.

The quantity of ground-up refuse used for fuel is only a small percentage (nominally 10% by heating value) of the fossil fuel normally fired to the boilers. However, the two test boilers fitted to burn refuse, although of only moderate size, can each consume about 300 tons of waste material per day at the 10% burning rate. This tonnage is equivalent to the residential solid waste generated by 170,000 people. If it is found feasible to burn larger percentages of refuse as fuel, the quantity of refuse disposed of by this means will increase accordingly.

Background

All major metropolitan areas within the United States face an increasing problem in the disposal of solid wastes. For the most part, the technology utilized up to this time for solid waste disposal has not been sufficiently sophisticated to permit efficient conservation of natural resources. Further, it is apparent that full advantage has not been taken of even proven areas of technology in the disposal of solid wastes.

The recovery of waste heat resulting from the combustion of solid waste materials is not a new concept and has been practiced in Europe for years. The practice has not been common in the United States; until recently it was confined to relatively inefficient waste heat boilers installed in conventional refuse incinerators. A few more sophisticated solid waste incinerators now are being built, incorporating boilers for the production of steam. These newer facilities unfortunately are quite costly and, to be effective as resource recovery units, must have markets for the steam they are designed to produce. Such markets are not always readily available.

Consider, however, a large, efficient, existing utility power plant, already integral with a system for producing, distributing and marketing electricity. It follows that if a means were devised to burn refuse as fuel in existing power plant boilers, without creating significant adverse effects, the matter of finding a market for the energy available from refuse would automatically be resolved, since utilities can market essentially all of the power they have the capability of producing.

The St. Louis project was conceived under the premises that if residential solid waste were properly prepared, and if it were to replace only a small percentage of the fuel fired to coal-fired boilers, the effects would be little, if any, different than if the fuel were entirely coal. Although coal-fired power plant boilers are not without operating problems of their own, a comprehensive study concluded that such problems would not be significantly increased, if increased at all, by burning prepared refuse as supplementary fuel. That study was conducted for the City by Horner & Shifrin, Inc. with the close cooperation of the Union Electric Company.

The concept was found attractive enough for the Union Electric Company to offer the use of two of its major power-producing boiler units for full-scale tests, and to contribute a substantial sum ($550,000) toward the installation of facilities to implement these tests. Also intrigued by the concept was the Office of Solid Waste Management Programs of the U.S. Environmental Protection

Agency, which partially funded the original study, and which, together with the Office of Air Programs, is partially funding the construction and operation of the facilities required for the full-scale demonstration program.

Process Details

Residential solid waste, in its raw state, is remarkably heterogeneous in appearance. After milling, however, the refuse becomes more homogeneous, in both appearance and consistency. Within reasonable limits, the analyses of refuse from different parts of the country have been found to be surprisingly consistent.

TABLE 2.1: RESIDENTIAL SOLID WASTE AND COAL SAMPLES (AS RECEIVED)

	Refuse * (%)	Coal + (%)
Proximate Analyses		
Moisture	19.69 - 31.33	6.20 - 10.23
Ash	9.43 - 26.83	9.73 - 10.83
Volatile	36.76 - 56.24	34.03 - 40.03
Fixed Carbon	0.61 - 14.64	42.03 - 45.14
Btu per Pound	4,171 - 5,501	11,258 - 11,931
Ultimate Analyses		
Moisture	19.69 - 31.33	6.20 - 10.23
Carbon	23.45 - 33.47	61.29 - 66.18
Hydrogen	3.38 - 4.72	4.49 - 5.58
Nitrogen	0.19 - 0.37	0.83 - 1.31
Chlorine	0.13 - 0.32	0.03 - 0.05
Sulfur	0.19 - 0.33	3.06 - 3.93
Ash	9.43 - 26.83	9.73 - 10.83
Oxygen	15.37 - 31.90	9.28 - 16.10

* From three samples of St. Louis residential solid waste, with magnetic metals removed.
\+ From three samples of Union Electric Company coals.

Investigations of the quality of residential solid waste produced in the St. Louis area, although limited in scope, disclosed characteristics similar to those found in other parts of the country. A comparison of the ranges of composition of refuse and coal for the St. Louis area was made (Table 2.1). As-received calorific values of refuse were found to be in the general range of 4,200 to 5,500 Btu per pound. The ash content was found to be about twice that of coal. Ash fusion temperatures were apparently very similar to those of Illinois bituminous coal. The sulfur content was found to be only a fraction of 1%. The chlorine content, however, was found to be higher than in most coals.

Processing Facilities

The solid waste to be processed for use as fuel, at least initially, will be confined to that produced from households. No bulky materials, such as appliances, furniture or tires, or wastes from industrial or commercial establishments will be processed. However, certain selected industrial and commercial wastes may be processed later.

Refuse processing, to produce 300 tons per day of supplementary fuel, will be accomplished during one 8-hour shift. Since magnetic metal will be removed, the equivalent quantity of raw refuse requiring processing will be about 320 to 330 tons per shift. Two-shift operation of the processing facilities will be required if the supplemental fuel is fired at rates above the nominal 300-tons-per-day rate. This will still permit the third shift to be devoted to scheduled maintenance of refuse grinding (hammermill) and other handling equipment. The hammermill and conveyors have been selected to provide a nominal production rate of 45 tons of raw refuse per hour.

Raw refuse is discharged from packer-type trucks to the floor of the raw-refuse-receiving building. Front-end loaders are used to push the raw refuse to a receiving conveyor. This method of handling raw refuse was selected over the pit and crane method in order to achieve more uniform feeding rates, to effect economy, and to provide an additional means of controlling the quality of the material being processed. From the receiving conveyor, the raw refuse is transferred to an inclined belt conveyor, which in turn discharges to a vibrating conveyor, which feeds the hammermill directly (Figure 2.1).

The hammermill is a conventional mill with a horizontal shaft. It has a hammer circle of about 60 inches, and an interior rotor length of about 80 inches. At the time the design of the mill was agreed upon, there was no evidence that any other type of mill could provide either the production rate required or the control over particle sizes desired for purposes of this process. The mill is powered by a direct-connected 1,250 horsepower, 900 rpm motor. Single-stage milling was deemed appropriate for purposes of the prototype installation, although two-stage milling normally is advocated for this type of operation. The mill has a grate cage which provides openings of about 2 inches by 3 inches.

Tests run on a similar but smaller mill indicated that essentially all milled particles would be less than 1½ inches in size, that 96 to 98% by weight of the particles would be less than 1 inch in size, and that about 50% of the particles would be less than 3/8 inch in size. Uncompacted bulk density of the milled material, depending upon moisture content and composition, was found to be variable, as low as 4 pounds per cubic foot in some cases, and as high as 12 pounds per cubic foot in others. This variation in density poses problems in equipment design and selection, since some equipment is designed on a gravimetric basis and others on a volumetric basis.

FIGURE 2.1: REFUSE PROCESSING FACILITIES

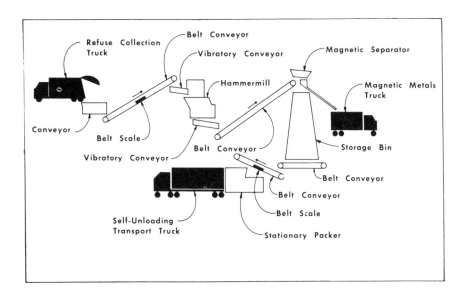

From the hammermill, the milled material is discharged to another vibrating conveyor, feeding an inclined belt conveyor leading to a storage bin. Magnetic separation is effected at the head pulley of this belt conveyor, with the magnetic metals discharged to trucks for sale as scrap.

The storage of milled refuse poses problems requiring special attention. For the prototype installation, since a convenient alternate means of refuse disposal was available (the processing plant was constructed adjacent to an existing city solid waste incinerator) and since the interruption of refuse-firing to the boilers would not cause significant operation problems, it was concluded that only minimum storage volume for the milled material was necessary. The storage volume provided, therefore, is only sufficient to permit a relatively even flow of supplementary fuel to the boilers.

Milled refuse, having laminar characteristics, has a bridging tendency, and storage bin design must be such that bridging will be minimized. The most effective means of preventing bridging appears to be the construction of bins with a greater cross section at the bottom than at the top. It also is necessary to provide a bin-unloading device which will remove material from all parts of the bin bottom without resorting to the use of hoppers, in which the material almost certainly would bridge. For this reason, traversing augers are being used to unload the storage bin.

From the storage bin, the supplementary fuel is conveyed to a stationary packer for loading into self-unloading trucks for transport to the power plant. In the prototype project, the power plant is located about 18 miles from the processing plant. At the nominal 300-ton-per-day firing rate, only one 25 ton load of supplementary fuel will be delivered to the power plant every two hours.

If it were feasible to locate the processing facilities near the power plant, it would be possible to eliminate truck transport by pneumatically conveying the supplementary fuel directly to the boilers from the storage bin.

Receiving and Firing Facilities

The self-unloading mechanisms of the transport trailers discharge the supplementary fuel to a receiving bin, from which the material is conveyed to a pneumatic feeder for transfer to a surge bin (Figure 2.2). The surge bin is equipped with four drag chain unloading conveyors, each of which feeds a pneumatic feeder. Each of these four pneumatic feeders conveys the supplementary fuel through a separate pipeline directly to a firing port in each corner of the boiler furnace. The pipelines are about 700 feet long.

The pneumatic systems are of the high-pressure type, in which the material to be conveyed is introduced by a rotary air lock feeder into a pressurized pipeline. The air velocities of the conveyed particles, depending upon their mass, are expected to be about 50 to 70 feet per second. The initial pressures in the pipelines depend upon their length as well as upon the quantity of material to be conveyed, and will normally be several pounds per square inch. The boilers are operated with balanced draft, with a slightly negative pressure in the furnaces.

The division of responsibility between the City of St. Louis and the Union Electric Company for this project was determined to be between the receiving facility and the surge bin. The operation of the surge bin and the pneumatic boiler firing systems have been established to be the responsibility of the utility. The delivery of the supplementary fuel and its transfer to the surge bin is the City's responsibility.

FIGURE 2.2: SUPPLEMENTARY FUEL RECEIVING AND FIRING FACILITIES

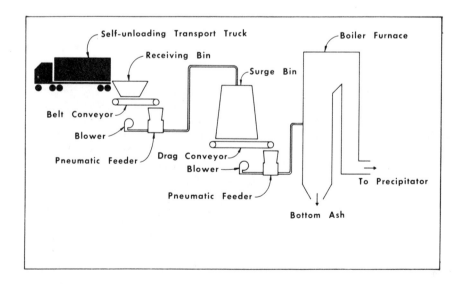

Test Boiler

The two twin boilers to be used for testing in the St. Louis project are small when compared to the newer units in the Union Electric Company system, but they are of modern, reheat design, and the test results from these units should be applicable to numerous other similar existing units in service in many parts of the country. Built by Combustion Engineering, Inc., whose personnel cooperated fully in the original study, the units each have a nominal rating of 125 megawatts, and each will burn about 56.5 tons of Illinois bituminous coal per hour at rated load.

FIGURE 2.3: MERAMEC UNIT NO. 1 UNION ELECTRIC COMPANY

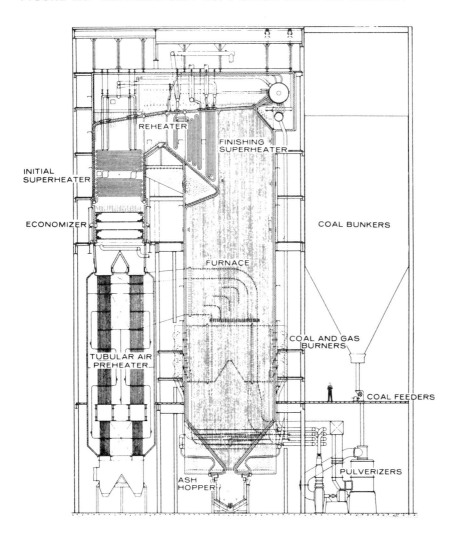

The units are tangentially fired, with four pulverized coal burners in each corner of the furnaces. Since they also are fitted to burn natural gas, they will be capable of burning refuse with gas as well as with coal during the test program. The furnaces are about 28 feet by 38 feet in cross section, with a total inside height of about 100 feet (Figure 2.3).

At full load, the quantity of refuse for each boiler equivalent in heating value to 10% of the coal is about 12.5 tons per hour, or 300 tons per 24-hour day. Refuse will be fired 24 hours per day, but only 5 days per week, since City residential solid waste collections are scheduled on a 5-day-per-week basis. No difficulty in boiler operation accruing from this interrupted refuse-firing schedule is anticipated.

Other than installing a refuse-burning port in each corner of the furnaces, no modifications to the test boilers were made. The refuse-burning ports were installed between the two middle coal burners. No alterations to the pressure parts of the boilers were necessary. The milled refuse is burned in suspension, in the same flame pattern as the pulverized coal or gas. As is typical of utility grade boilers, the furnaces have no grates. Nonburnable particles, or burnable particles with sufficient mass to prevent them from being consumed in suspension, therefore are likely to fall to the bottom ash hopper.

The prepared refuse is fired at a constant rate. The existing boiler combustion controls vary the rate of firing of the pulverized coal or gas to accommodate the heat requirements of the boilers. Should the boiler go out of service suddenly, for any reason, an electrical interlock will immediately stop the feeding of refuse, although the pneumatic blowers will continue to function in order to clear the pipelines of any refuse remaining in them.

No modifications were made to either the electrostatic precipitators or the bottom-ash-handling systems. Although a significant increase in the quantities of ash was anticipated, particularly in bottom ash, the existing ash handling facilities were regarded as adequate, at least for purposes of the test. Since both fly ash and bottom ash are marketed, tests will be conducted to ascertain changes in ash quality as a result of burning refuse as supplementary fuel. However, the low refuse-burning rate is not anticipated to substantially change the ash characteristics.

Combustion Engineering, Inc., in their initial appraisal of the applicability of the process to the test boilers, concluded that the air and gas weights would be very close to the original design conditions, and that the forced draft and induced draft fans would be adequate.

Potential Boiler Operating Problems

Although the limitation of burning residential solid waste as only a small percentage of the total fuel should tend to minimize any potential additional boiler operating problems, there are several potential effects which are being given special attention in the tests. Slagging, corrosion, completeness of burn-out, and precipitator performance are, therefore, being thoroughly evaluated.

Some slagging can be expected in most coal-burning boilers, with the degree of slagging related to coal quality and to operating procedures. Preliminary investigations indicated that the ash fusion temperatures of refuse were in the same range as those of Illinois bituminous coal. Other coals, particularly those with low sulfur content, may have higher ash fusion temperatures. The tests will include observations of potential slagging tendencies of mixtures of refuse ash and coal ash.

An evaluation made by Combustion Engineering, Inc., during the preliminary studies indicated the possibility of a slightly increased corrosion potential. Probes will be installed at strategic points in the boiler to assess the effects, if any, of corrosion. The removal of magnetic metals, along with the zinc, lead, and tin that often will be bonded to them, is expected to reduce the potential adverse effect of corrosion and plating caused by metal oxides.

Since utility boilers are not normally equipped with grates, there was some concern that larger burnable particles would not be completely consumed in suspension. The small particles achieved in the type of milling equipment provided is expected to decrease this possibility, particularly in view of the large percentage of paper in refuse.

When burning pulverized coal and milled refuse in combination, some increased particulate loading may occur in the boiler exhaust gases which are cleaned by an electrostatic precipitator. Therefore, tests will be conducted to determine the relative precipitator performance.

Air pollution tests will also be conducted on the stack gases to evaluate the effect of refuse combustion upon gaseous pollutant emissions. It is expected that the firing of refuse and coal will not produce discernible increases in either gaseous or particulate emissions when compared to burning coal alone.

Economics

Of particular importance are the relative economics of the application. The basic economic elements are: (1) the total unit cost of processing, transporting, and firing the refuse as supplementary fuel, including those costs accruing to the utility; (2) the costs of alternate means of residential solid waste disposal; (3) the value of the magnetic materials recovered; (4) the value of the milled residential solid waste as fuel when compared to the value of fossil fuel. Under almost any situation, the capital cost of the facilities required for application of the process would be substantially less than that of conventional incineration facilities. Total unit costs per ton, including operation, maintenance, and amortization, could be comparable to those of sanitary landfill when the landfill is in a location remote from the point of refuse generation.

Utility fuel costs are rising rapidly, and even under present conditions the relative value of refuse as fuel could be significant. The net cost of residential solid waste disposal by this means, therefore, could be highly attractive in many of the metropolitan areas of the United States.

Applicability of Process

The prototype installation for burning residential solid waste as supplementary fuel includes the adaptation of tangentially-fired boilers. Indications are that such units are the most easily adaptable of the various types in common use. However, there appears to be no reason why cyclone-fired boilers could not be adapted, or why front-fired and opposed-fired units could not be fitted with refuse-burning ports, even if some modification of pressure parts were to be required.

Although the original investigations were confined to pulverized coal-burning boilers, the test boilers are fitted to burn natural gas as well, and therefore are capable of burning solid waste with gas as well as with coal during the testing program. There is no reason why boilers burning oil also could not be adapted to burn solid waste, provided they have the capability of handling both bottom ash and fly ash.

The capability of existing suspension-fired boilers to consume municipal solid waste as supplementary fuel is great enough to permit the process to serve as a principal means of solid waste disposal for many metropolitan areas, even when the supplementary fuel is fired as only 10% of the total fuel requirement. A 600-megawatt unit, for example, could at full load consume on the order of 1,200 tons of solid waste per day at the 10% rate. It was estimated that by 1973 the suspension-fired boilers in the St. Louis area will have the potential capability of consuming over twice the amount of municipal solid waste produced in the entire metropolitan area with a population of over 2,500,000.

The process cannot be expected to be applicable in all metropolitan areas. Some types of fossil-fuel-fired boilers are not as adaptable as others. Boilers operating essentially as base load units may be more desirable for burning solid waste than those operated at partial loadings. Cooperation between utilities and municipal corporations sometimes may be difficult to achieve. The quantities of supplementary fuel available in some areas may not be sufficient to be of interest to the utility. The location of the power plant may not always permit economical application of the process.

In the case of the prototype project, the City of St. Louis is delivering the prepared residential solid waste to the utility at no cost for the duration of the tests. Depending upon the circumstances of a given case, this same arrangement could be mutually attractive to another combination of a city and a utility, or a city might be willing to pay the utility a nominal fee for using the solid waste, or the utility might be willing to pay the city a nominal fee for the supplementary fuel. Each situation requires an objective appraisal to determine the appropriate basis for negotiation.

The process shows promise of having fairly extensive application once a satisfactory financial arrangement can be worked out between the utility and the solid waste supplier, whether it be a governmental unit or a private organization. It could provide an economical means of disposing of large quantities of solid waste, effect recycling by the direct recovery of electrical energy, conserve natural resources of fossil fuel and magnetic metal, and, to some degree, assist in the control of air pollution. The preliminary appraisals indicate that these potential benefits may be achieved by effecting sufficient cooperation between municipalities and utilities to permit use of existing large-scale power plant facilities, by using commercially available equipment and by applying already tried and proven basic technology.

Lynn, Mass.
Refuse-Fired Waterwall Incinerator Study

On October 9, 1967, the major dumping area of Greater Lynn, the DeMatteo landfill, was closed as a result of action taken by a group of citizens who were dissatisfied with the operation of the facility. Though a "stay of execution" had been granted by the Commonwealth of Massachusetts, allowing selected communities to continue using the facility, the problem of refuse disposal in this area has become critical. It is clear that continued use of the DeMatteo area will not be tolerated much longer.

The City of Lynn, excluded from using the DeMatteo facility, formulated an agreement with the New England Power Company whereby it would establish a sanitary landfill on an area of land owned by that firm. This is the only current source of land for refuse disposal in the City of Lynn and it is rapidly being depleted. A plan to construct an adequate solid wastes facility is absolutely necessary.

The General Electric Company's River Works Plant, located in Lynn, has a need for additional steam. This need for steam at the industrial facility, and the severe problem of disposal of solid wastes in the community, have brought the General Electric Company and the City of Lynn together in this joint venture.

There are facilities in operation, such as the United States Navy's incinerator-boiler at Norfolk, Virginia, which utilize the heat from the combustion of refuse for steam generation. It was therefore proposed that the two parties combine in a joint venture for their mutual benefit. The refuse will have a value as a fuel to the General Electric Company, and the disposal service will have a value to the community.

Application was made to the Bureau of Solid Wastes Management of the United States Public Health Service for a study and investigation grant to help finance a study of the concept of utilizing refuse from the City of Lynn as a low-grade fuel to generate steam for use by the General Electric Company, thereby achieving a mutual economic savings. Metcalf & Eddy, Inc., Boston, Massachusetts, and the Foster Wheeler Corporation, New York, New York, were engaged by the City of Lynn and the General Electric Company, respectively, as consultants.

The study investigated several alternative solutions with varying degrees of involvement, in regard both to initial capital outlay and to the operation of the facility by the two parties. The recommendations apply to a situation existing in the City of Lynn, Massachusetts, and is based on several conditions that apply only to the study area. It is felt, however, that the basic concept is applicable to any location in the United States where there is a market for steam. For the recommendations to be applicable to other localities, it would be necessary to obtain specific data for the situation and reanalyze. The published report compared the economic feasibility of three alternatives for the ownership and operation of the proposed steam-generating system:

(A) Waste-preparation plant to be owned and operated by the City of Lynn; the boiler and superheater to be owned and operated by the General Electric Company. (There would be 800 foot long conveyors connecting the two facilities.)

(B) Waste-preparation plant and boiler to be owned and operated by the City of Lynn; the superheater to be owned and operated by the General Electric Company. (There would be an 800 foot long steam main between the boiler and the superheater.)

(C) Same ownership and operation as alternative (A), but the two plants would be located adjacent to each other; therefore, the 800 foot long conveyors would not be necessary.

BASIC DESIGN CRITERIA

Heating Value of Refuse: The heating value of a mixture consisting of several separate substances can be determined by using a weighted percentage based on the weight of each separate substance. Because the composition of refuse varies greatly, and because many different substances with differing heating values are found in refuse, the heating value has been found to vary greatly. For example, commercial refuse, because of the relatively high proportion of garbage, would have a lower heating value than industrial refuse, which has little garbage in it. The heating values used in this report are shown in Table 3.1.

TABLE 3.1: PRESENT HEATING VALUES OF REFUSE

Source	Heating Value, Btu/Lb. as Fired
Residential refuse	5,000
Commercial refuse	4,000
Industrial refuse	6,000

The composite heating value of refuse that would be expected at the process plant was determined by multiplying the weight of residential, commercial, and industrial refuse by the respective heating values shown above, and then dividing by the combined weight of all the refuse. The resulting heating value was found to be 4,970 Btu/lb. as-fired. The trend in the past has been a steady rise in heating value due to the decreased proportion of garbage and increased proportions of combustible packaging materials in the total refuse. Based on this trend, a heating value of 6,000 Btu/lb. as-fired was assumed for the year 1990.

General Electric's Steam Requirements: The General Electric Company has requested that steam be provided at boiler outlet conditions of 650 psig and at a temperature of 850°F. The demand is not constant over a 24 hour period, but varies from a minimum of 200,000 lbs./hr. to a maximum of 400,000 lbs./hr. Hourly rates, together with the percentage of time they are required, as established by General Electric, are shown in Table 3.2. The steam requirements in Table 3.2 include an allowance for the steam required to preheat the feedwater and to operate the deaerator. Accordingly, even if the boiler is to be operated by the City of Lynn, the total steam production required from the boiler would remain as shown in the table.

TABLE 3.2: GENERAL ELECTRIC'S STEAM REQUIREMENTS

Demand, lbs./hr.	Percent of Time
400,000	5
350,000	10
300,000	50
200,000	35
Total	100

Boiler Requirements: The boiler must be capable of handling over 300 tons of refuse per 24 hour day. Ability of the equipment to handle larger quantities of refuse is a desirable feature. The physical size of the boiler is, however, limited by the space available in the General Electric Company yard. The boiler will be capable of generating the full load (400,000 lbs./hr.) when firing a combination of oil and refuse and also when firing oil alone. It will also be capable of firing over the full range indicated in Table 3.2. General Electric has established requirements that the boiler be entirely above ground level and that the firing aisle be 20 feet above the present ground floor elevations.

Backup Facilities: The design of the processing plant will be such that the operating capacity is obtainable with any one piece of equipment out of service. This backup may be provided with standby equipment, by operating other equipment for additional hours during the day, or a combination.

Available Utilities: The existing fuel storage and distribution system, and portions of the existing boiler feedwater treatment facilities, at the General Electric Company would be used for the new boiler. The General Electric Company generates its own power which will be used to operate both the equipment associated with the boiler and the refuse-processing equipment. Water for drinking and fire protection can be obtained from a Metropolitan District Commission water main located on Route 107, which is about 1,000 feet from the proposed plant, or from the General Electric Company distribution system. Sanitary sewage originating on the southwest side of the Saugus River can be collected and pumped across Route 107 to an existing sanitary sewerage system.

Refuse Densities: The density of refuse varies depending upon the degree of compaction and the particular constituents of the batch being investigated. For instance, municipal refuse at the curb side has an average density between 5 and 10 pcf (pounds per cubic foot). When placed in a packer truck, the density reaches about 20 pcf. Refuse compacted in huge compaction machines can reach a density of over 60 pcf. The densities used in this report are presented in Table 3.3.

TABLE 3.3: REFUSE DENSITIES

Degree of Preparation	Pounds/Cubic Foot
Normal refuse	
Packer truck	18.0
Storage bin	12.5
Shredded refuse	
Conveyor	6.0
Storage silo	15.0
Residue	
Dry	40.0
Wet	50.0

Refuse Generation: The estimated residential commercial and industrial refuse generation rates are summarized in Table 3.4. Predictions for the next 20 years vary from ½ to 5% increase per year. We have assumed an average value of 2% per year. The 1990 estimated residential, commercial, and industrial refuse generation rates based on an average annual increase of 2% are shown in Table 3.4.

The residential refuse production in the City of Lynn, together with General Electric's industrial contribution, is estimated to be 176.5 tpd in 1970 and 262.5 tpd in 1990. Two alternative refuse-burning systems were proposed by Foster Wheeler. The first alternative incorporates a reciprocating grate stoker; the second, a spreader stoker. The boiler using a reciprocating grate stoker has a capacity of 384 tpd; the one using a spreader stoker has a capacity of 612 tpd.

TABLE 3.4: ESTIMATED REFUSE PRODUCTION IN LYNN

Refuse Classification	Tons per Day*	
	1970	1990
Residential	160.0	238.0
General Electric (industrial)	16.5	24.5
Subtotal	176.5	262.5
Commercial	45.0	67.0
Other industrial	16.5	24.5
Total	238.0	354.0

*Based on a 7-day week.

Lynn, Mass.: Refuse-Fired Waterwall Incinerator Study

BASIC DESCRIPTION

The report analyzed a number of alternatives. The most interesting for the purposes of this book was having both the preparation plant and the boiler located together in Saugus, utilizing a spreader stoker with complete shredding.

The process plant and the boiler with spreader-type stoker would be owned and operated by the City of Lynn and located on the south side of the Saugus River. Saturated steam will be conveyed to the General Electric Company, where it will be superheated in a separately fired superheater, owned and operated by the General Electric Company. A site plan is shown in Figure 3.1. A flow diagram and layout are shown in Figures 3.2 and 3.3. Trucks carrying all types of refuse will be weighed on the platform scale prior to entering the process plant. The trucks will then proceed across the dumping floor and empty into the storage bin. The refuse storage bin will have a capacity for 1.4 days' storage, based on a stoker rate of 610 tpd.

FIGURE 3.1: SITE PLAN

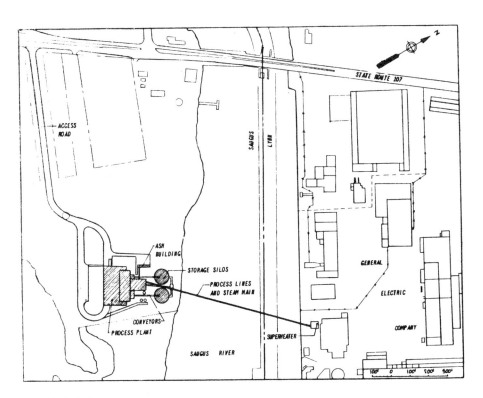

SITE PLAN

FIGURE 3.2: SCHEMATIC FLOW DIAGRAM

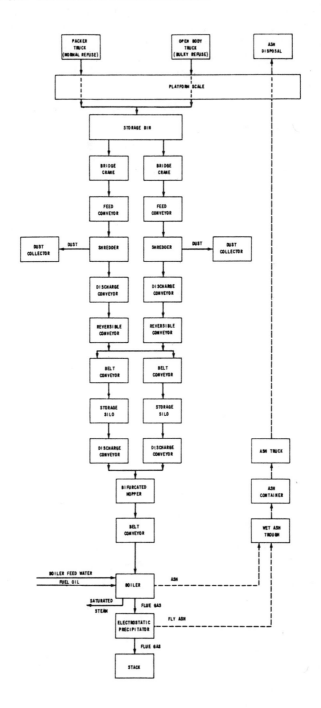

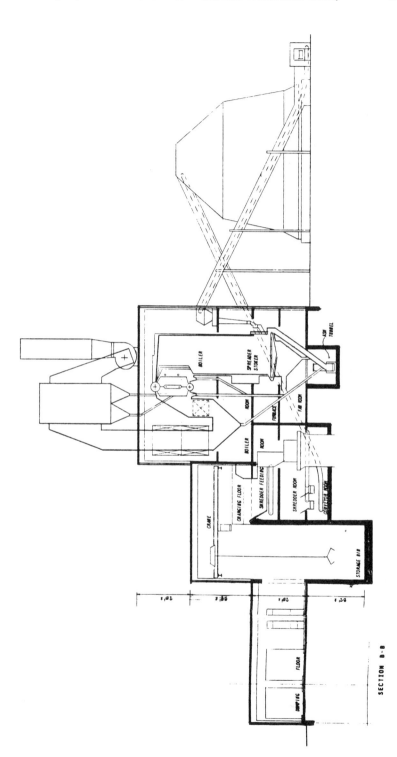

FIGURE 3.3: LAYOUT

The refuse will be removed from the storage bin, placed on conveyors, shredded, conveyed into silos, and metered from the silos. The silos will discharge onto a series of conveyors which will convey the material up to the boiler. The boiler will produce saturated steam. This saturated steam will be conveyed approximately 800 feet across the Saugus River to the superheater located in the General Electric yard. Here the steam will be superheated and put into the General Electric distribution system. All of the utilities (boiler feedwater, fuel oil and power) will be obtained from the existing facilities at the General Electric Company. The residue from the boiler and the precipitator will be collected and quenched in a water-filled trough beneath the boiler and conveyed to containers located in the ash building.

UNITS OF REFUSE PROCESSING SYSTEM

Weighing Facilities: Maintaining records of the amount of refuse entering an incinerator and the amount of ash leaving the incinerator serve as an excellent source of information for checking incinerator operation. At incineration plants that serve more than one community, charges can be apportioned on the basis of the weight of the incoming refuse.

Scales are available which will automatically record the weight of the refuse vehicle on a card inserted into a recorder attached to the scale. Since communities other than Lynn may be using the incinerator, the scale should be long enough to accommodate the largest refuse collection vehicles presently on the market. This appears to be an 80 cubic yard transfer vehicle which has an overall length of over 50 feet. The scale should also be of sufficient capacity to handle the heaviest vehicles allowed on the highways. Based on these criteria, the scale should be 60 feet long, with a capacity of 40 tons.

Once a vehicle's tare weight has been established, this can be used each time the vehicle returns to the incinerator with only an occasional check. Therefore, only the gross weight need be recorded each time the truck enters. Subtracting the tare weight from the gross weight gives the weight of refuse on the vehicle. The weighmaster is located so that he can easily see both the truck being weighed and the dumping floor. He can then direct the drivers to the appropriate section of the dumping floor.

Dumping Floor: Due to the climate and for aesthetic reasons, most of the larger incinerators constructed in New England have an enclosed floor where the refuse vehicles maneuver prior to dumping their refuse. The dumping floor should extend the full length of the refuse storage area and should be of sufficient width to accommodate the largest refuse vehicle that could be expected to use the facility. The length is also governed by the number of vehicles expected to be dumping at one time.

There should be dumping space for a minimum of 10 and 15 vehicles, respectively, under the 384 and 612 tpd refuse-burning systems. A width of 80 feet would provide sufficient maneuvering space for the largest transfer vehicles in use today. In the alternatives where only the bulky refuse is shredded, a space approximately 50 feet wide by 75 feet long has been reserved on the dumping floor for the storage of bulky objects.

Storage Bin: General Electric's steam requirements are such that the boiler must be operated 24 hours per day, 7 days per week. Refuse would be delivered to the plant Monday through Friday, during the day only. Therefore, some means of refuse storage will be required. Refuse may be stored in silos only if it is shredded. The volume of the storage bin is calculated using a density of 333 lbs. per cubic yard for refuse in the bin. The width of the bin has been set at 28 feet. The depth of the bin is influenced by the condition of the underlying soils. There is approximately 35 feet of silty material underlain by several layers of clay. The clay appears as if it would support the bin without piles, whereas piles would be required if the bin were placed in the silt layer. Therefore, the bin was made deep enough so that the base would rest in the clay layer. With the depth and width established, the length was calculated from the total required volume.

Where all material is shredded, it is assumed that the entire volume of refuse is delivered to the storage bin in approximately equal portions Monday through Friday, and that all of the refuse is shredded and delivered to the storage silo during the same day that it is delivered. A storage bin capacity equal to the quantity to be delivered in one day was selected. This capacity will provide time for maintenance without interfering with the boiler operation, should some part of the processing system break down. The bin capacity will be seven-fifths of the daily burning rate or 1.4 days' storage.

Cranes: Based on the size of the storage bins, and the quantities of refuse to be handled, bridge cranes were deemed to be the best means to transfer the refuse from the storage bin to the charging hopper. The crane sizes were selected based on the distances the crane would have to travel, speeds recommended by the manufacturers, an effective working time of between 40 and 50 minutes per hour, and the quantity of refuse required by the downstream capacity of the system. A minimum of two cranes have been provided.

Shredders: Employing a spreader stoker, the refuse will be shredded to a maximum of 4 inches in the longest dimension. In the alternatives employing a reciprocating grate stoker, only the bulky refuse would need to be shredded. Several makes of shredders were investigated including both the horizontal and vertical shaft models. The size of the shredder is somewhat governed by the rate at which it can be fed. The preliminary building designs have sufficient space for either the vertical shaft or the horizontal shaft shredders.

Based on one crane feeding one shredder, the maximum practical quantity that a crane could handle would be between 30 and 40 tph. Therefore, shredders with a capacity in this range were selected. The number of machines and hours of operation were then determined. Because shredders are a high maintenance item, backup was provided. Two shredders have been provided where total shredding is employed, since a major overhaul would otherwise shut down the system. Minor repairs and routine maintenance could be performed during the hours that the shredder is not operating. Where only bulky refuse is shredded, the size of the shredder is based on the feed opening required to accept bulky material such as mattresses, packing crates, furniture, etc. The bulky refuse shredders would discharge onto a conveyor emptying directly into the storage bin.

Dust Collector: As the shredder operates, the speed of the rotor causes air to flow through the machine. When material is being shredded, this exhausted air is loaded with dust. Therefore, the air is passed through a scrubbing type dust collector before it is discharged to the atmosphere. The particulate material collected in the dust collector is discharged onto the conveyor system downstream of the shredder or into the storage bin.

Conveyors: Rubber belt conveyors were selected in all cases except where cranes, front end loaders or shredders would discharge onto them. Under such circumstances, belt conveyors would experience an excessive degree of wear; therefore, metal pan conveyors were selected instead. Where refuse is to be discharged from another belt, the rubber belt type was selected for the receiving conveyor. The initial cost of metal pan conveyors is about twice as much as rubber belt conveyors; however, the savings in belt replacement and downtime more than offsets the initial cost differential. Reliability is provided by using reversible belts and in other instances parallel belts. The angle of incline of all conveyors transporting refuse is kept to a maximum of 27 degrees.

The width of the conveyors is selected considering a variety of factors. The width of the infeed conveyors to the shredders, and the discharge conveyors from the shredders, was based on the physical dimensions of the shredder. The width of the conveyors fed by the cranes, but not discharging to shredders, was based on the physical dimensions of the crane bucket. Conveyors carrying shredded material were sized by the volume of material, the density of the material, and the speed of the conveyor. The width of the conveyors carrying unshredded refuse across the Saugus River was set at 10 feet. The width is necessary, not for the volume of material, but because large items which inadvertently bypass the shredder may become lodged and cause a backup if a narrower width were selected.

The conveyor handling ash from the boiler will be a drag-type conveyor. This type of conveyor has metal flights suspended from a traveling chain operating in a water filled trough. The residue will be discharged from the boiler into the water filled trough. As it settles to the bottom of the trough, the metal flights drag the ash along the bottom of the trough and up an incline where most of the water is drained from the ash. The water filled trough also serves as a seal on the boiler to prevent hot gases from escaping. Pneumatic conveyors were considered for conveying shredded refuse but they were found to be much more expensive overall than mechanical conveyors, and therefore are not recommended.

Storage Silos: Storage silos can be discharged automatically; therefore, they provide a much more economical means of storage than a bin which requires a bridge crane and operator to empty. Silos, however, can only be used to store shredded refuse; therefore, they are only used where all of the refuse is shredded. With storage silos, it would be possible to shred the total weekly volume of refuse during one or two 8 hour shifts per day under the 384 and 612 tpd alternatives, respectively. This would place the majority of the labor required for the process operation on a 40 hour week and would reduce the total number of personnel. The silos would have a positive discharge, which is necessary because of the unpredictable nature of the material. The bottom diameter of the silo is such that bridging will not occur.

The refuse is not expected to freeze in the silo. Some freezing may occur near the outer edge of the silo, but this should act as an insulation to prevent further freezing. It would be possible to insulate and heat the silos at a later date if freezing is found to be a problem. The capacities of the silos were based on the amount of refuse needed in the boiler from the end of the last shift on Friday evening until the beginning of the first shift on Monday morning, plus several hours' backlog. Two silos would be used in all instances. Each silo would have a capacity of one-half the total requirement. Each would be capable of discharging at a rate equal to the capacity of the boiler so that, in the event one silo is down for repairs, the overall operation is not affected. Based on the above criteria, two silos, each with a capacity of 650 tons, would be needed for the 384 tpd alternatives, and two silos, each with a capacity of 910 tons, would be needed for the 612 tpd alternatives. The rate of discharge from a silo can be varied over a wide range, and the silos can therefore be used to meter the quantity of refuse fed into the boiler.

Bulky Refuse Handling: Bulky refuse is composed of large items such as bureaus, upholstered chairs, sofas, mattresses, tree trunks, washing machines, refrigerators, etc., discarded from the home; and pallets, packing crates, machinery parts, and other large items discarded from industrial and commercial establishments. Bulky refuse can be subdivided into two categories: combustible and noncombustible. In this report, combustible refuse is material that will burn at temperatures equal to or less than the normal furnace operating range of 1,600° to 2,000°F. Noncombustible bulky refuse will not be accepted at the plant.

In the alternatives where refuse is fired on a spreader stoker, all of the refuse will be shredded. In these alternatives, combustible bulky refuse will be deposited into the storage bin and follow the same flow pattern as the remainder of the refuse. In the alternatives where refuse is fired on a reciprocating grate stoker, the refuse will be fired as it is received, except that Foster Wheeler requires that bulky items be reduced in size prior to firing. Therefore, a shredder is provided in these alternatives for bulky refuse.

Bulky items are placed directly on the dumping floor. From here a front end loader places the bulky items on an apron conveyor feeding the shredder. A short apron conveyor underneath the shredder discharges the shredded material directly into the storage bin. Since the combustible bulky refuse is composed mainly of wood, which has a relatively high heat value compared to average refuse, the bridge crane will be used to mix the shredded bulky refuse with the remainder of the refuse prior to charging, to avoid uneven Btu loading on the boiler.

Residue Handling: The residue from an incinerator can be divided into three general categories:

> *Bottom Ash* — Bottom ash is the noncombustible and unburned material remaining on the grates. The bottom ash will be discharged from the end of the grate directly into a water filled trough.
>
> *Fly Ash* — Fly ash is the small particulate matter collected in the electrostatic precipitator. Where the electrostatic precipitator is installed at ground level, the fly ash will be discharged onto a slider-belt conveyor and conveyed to the water trough. This material is very fine in nature, ranging in diameter from 120 to less than 5 microns; therefore, a steam spray will be installed to wet the material down to prevent it from blowing about. Where the electrostatic precipitator is installed on the roof of the building, the

fly ash will be conveyed by gravity in pipes down to the water trough.

Miscellaneous Residue — The siftings collected under the grates, and particulate matter collected in miscellaneous hoppers throughout the process, constitute the miscellaneous residue. This residue will also be conveyed to the water trough.

A drag conveyor in the bottom of the water-filled trough will move the residue from the trough up an incline where most of the water will be drained from the residue. The residue would go directly from the drag conveyor into 30 cubic yard ash hoppers. The quantity of ash, on a dry basis, is estimated to be 20%, by weight, of the original quantity of refuse. The density of dry ash is approximately 40 lbs./cu. ft. and the density of wet ash is approximately 50 lbs./cu. ft., and the average moisture content is 20%. Therefore, the average quantity of ash from the 384 tpd reciprocating grate fired alternatives would be approximately 100 tpd, and the quantity of ash from the 612 tpd spreader stoker fired alternatives would be approximately 150 tpd. These are equivalent to an annual landfill volume of about 30 and 50 acre-feet, respectively.

Incinerator residue has been used successfully as a landfill material to reclaim unused tracts of land. Areas where groundwater would pass through the residue should be avoided, however, since soluble salts and alkalies would be leached from the ash and carried along with the groundwater. The residue anticipated from the boilers considered in this report should have a relatively small percentage of volatile matter and therefore should be relatively stable. The effective use of residue to reclaim land areas, thereby increasing the city's tax base, can help to offset the cost of residue disposal.

Under the reciprocating grate fired alternatives, the fly ash would be approximately 25% of the total residue, by weight. Under the spreader stoker fired alternatives, because of the small particle size and the nature of the firing, the fly ash would be approximately 50% of the total residue, by weight. However, since all of the residue would be transported to the same water-filled trough, the method of refuse handling as outlined above would be applicable to either type of firing. The weight of all residue leaving the plant should be recorded. This can easily be accomplished by weighing the residue trucks on the same platform scale used to weigh the incoming refuse collection vehicles. This weight when compared to the weight of refuse entering the plant can be used to check on the efficiency of the operation.

UNITS OF REFUSE BURNING SYSTEM

Steam Generating Unit: The refuse burning, steam generating unit will be a natural circulation, water-walled boiler with a stoker specifically designed for refuse burning. When operating with refuse having an as-fired heating value of 4,970 Btu/lb., the boiler will have the capacity to burn 384 tpd with a reciprocating stoker, and 612 tpd with a spreader stoker. The refuse burning capacity of the boiler with 6,000 Btu/lb. of refuse will remain at 384 tpd for the boiler with the reciprocating grate stoker. The boiler capacity with 6,000 Btu/lb. of refuse and a spreader stoker will decrease to 510 tpd, since this grate is designed on the basis of 750,000 Btu/sq. ft. regardless of the refuse characteristics.

The boiler burning refuse and No. 6 oil simultaneously will produce 400,000 lbs. of saturated steam at approximately 750 psig, which will, in turn, be delivered to a separately oil-fired superheater. The superheater is designed to deliver 650 psig steam at 850°F. into the General Electric Company distribution system. Normally, the superheater is an integral part of the boiler. Because of the corrosive nature of refuse gases containing constituents such as chlorine, the superheater will be separated from the incinerator and thus will prevent impingement of these gases on the high temperature tubes of the superheater. Although the separately fired superheater is located on General Electric Company property in all alternatives, it would be possible to locate the superheater adjacent to the boiler on the Saugus side of the river if advantageous to do so. Location of the superheater on General Electric Company property is desirable, however, from a heat loss standpoint, since the thermal gradient or heat loss driving force is about 1.5 times as great for the superheated steam as it is for the saturated steam due to the higher temperature of the superheated steam.

At the full load design rating, the boiler will be designed for a range of 40 to 100% excess air for the refuse component of the fuel, and approximately 15% excess air for the No. 6 oil component. When using the spreader stoker and burning 612 tons of refuse per day and sufficient oil to produce 400,000 lbs. of steam per hour, the calculated overall boiler efficiency will be 78.5%. When using the reciprocating grate stoker and burning 384 tons of refuse per day and sufficient oil to achieve the same steaming rate, the boiler overall efficiency will be 81.4%. On oil alone, for the same steaming rate, the boiler overall efficiency will increase to 86.4%.

Reciprocating Grate Stoker: The reciprocating grate stoker is designed in lateral rows, each overlapping its upstream neighbor in a shingle-like manner. Alternate rows are linked to a hydraulic power cylinder which slowly reciprocates them back and forth across the face of the alternate stationary rows.

The proposed unit for this application will consist of four independently driven and controlled stoker sections. The first section will be a charging section and the three remaining sections will be designed for combustion of the refuse. A vertical drop-off is provided between grate sections to break up and reorient the refuse to provide maximum surface exposure to the flame.

Pressure drop through the grate venturi air openings is relatively high to ensure a more uniform air distribution and to minimize the effects from the varying characteristics of the refuse bed. Not only does this provide more uniform combustion, but it also ensures more uniform cooling of the grates. Feed is uniform and continuous, and noncombustibles and ash are discharged continuously from the grates. Siftings removal is automatic. The removal devices are tied to the stoker. The siftings are conveyed to the discharge end of the unit and discharged along with the bottom ash to the ash conveyor.

Spreader Stoker: The spreader stoker system consists of a traveling grate, four air-swept spouts and a swinging distributor assembly. The grates are divided into rows longitudinally, one row for each air-swept spout or feeder. Each row is carried on two chains which ride over hardened tooth sprockets. A hydraulic system drives the grate toward the front of the boiler, discharging the ash immediately below the feed spout. The grate design is specifically for suspension firing, and individual grate bars open on the return run of the grate facilitating air admission to the fuel bed, and the discharge of siftings to the chamber below the grate. The air-swept spouts spread the shredded material evenly over the grate. The lighter material burns in suspension and the heavier material falls to the relatively fast moving grate. The air to the spout is controlled by a motorized rotary air damper, which alternately increases and decreases both quantity and pressure of the air several cycles per minute. This assures even fuel distribution from front to rear of the furnace.

Control System: The steam, condensate and combustion control systems will be pneumatic, or combination pneumatic and electric operated. With the water-walled incinerator in Saugus, feed control will be achieved by stoker speed variation only. In both cases, a bridge crane will be utilized to move the material from the storage pit to the conveyor or incinerator charging hopper.

With total shredding, feed control will be achieved by varying the silo discharge and associated conveyor speeds in conjunction with the stoker speed, regardless of the water- walled incinerator location. It should be noted that the refuse feed control should be minimal since the anticipated steam requirements exceed that which can be produced by the refuse alone. In addition to the refuse feed control, the fuel oil system will have modulating flow rate controls. This system will meter the fuel as the required steam flow changes over the anticipated range. This, together with automatic control of the separately fired superheater, will ensure that steam characteristics are well maintained. Regardless of the water-walled incinerator location, the feedwater supply to the incinerator will be controlled by a three-element controller, i.e., steam flow, feedwater flow, and drum level.

The feedwater supply to the boiler will in all cases originate from the General Electric Company property. Their softening, polishing and storage facilities must be expanded to accept the added load, however. A condensate tank will be required adjacent to the incinerator. Softened water will be pumped from General Electric to this tank and from this tank to the deaerator. The condensate storage tank will have at least 20 minutes' storage capacity in addition to the 10 minute storage in the deaerator to ensure safe shutdown of the incinerator upon inadvertent loss of water. A steam-turbine drive will be provided on one of the condensate pumps at the storage tank and on one of the boiler feed pumps to maintain feedwater to the boiler in the event electric power is lost.

A separate storage tank adjacent to the boiler will be necessary as intermediate storage for No. 6 oil. The level in this tank will be controlled by a float switch connected to an electrically operated control valve. The valve will be located at the inlet to the tank. The No. 6 oil tank will be large enough to hold approximately 20,000 gallons of oil to provide an alternate source of oil should the supply originating from General Electric be interrupted. The fuel oil system for the boiler will be controlled by steam pressure and for the superheater by temperature. Both these systems would be designed to meet local and state safety codes.

Water-Softening System: The water-softening system will have sufficient capacity to process all raw water to the boiler assuming 100% makeup, and also a polishing system to remove primarily iron and copper from the recycled condensate. This condensate will be pumped in plastic lined steel pipe either directly to the deaerator or via the fiber glass condensate tank, to the deaerator, depending upon the location of the water-walled incinerator.

A predetermined level will be maintained in the condensate storage tank by a float operated control valve. Sufficient heat will be supplied to the tank to prevent freezing when the system is not in operation. The pipe crossing the river will be on roller supports to allow for expansion and will have expansion joints at both ends. A minimum blow bypass valve will be provided downstream of the condensate pumps to allow the condensate to flow back to the tank under low-flow conditions and prevent damage to the pump due to overheating.

Boiler Feed and Steam Systems: Downstream of the deaerator, the feedwater will flow to two boiler feed pumps, one motor driven and the other steam turbine driven. Both pumps will be sized to take the full boiler capacity. A recirculating valve will be provided on the discharge of each pump. At excessively low flow rates the feedwater will be bypassed to the deaerator storage tank. The boiler feed pump drive turbine is designed for 200 psig steam. Its exhaust will be piped to the deaerator which operates at 5 psig. In order to obtain the necessary feedwater temperature for the boiler studied, a feedwater heater is provided between the boiler feed pump and the economizer. Steam from the main heater will be reduced to approximately 175 psig and piped to the feedwater heater. The condensate from the feedwater heater will then go directly to the deaerator. After having been heated to 365°F. in the feedwater heater, the feedwater will be pumped to the economizer and then to the boiler itself.

The steam pipe crossing the river will rest on roller supports to provide a means of expansion. The pipe will be suspended about 50 feet above the Saugus River and the boiler and superheater connections will be about 50 feet above this elevation. With this configuration four hinged expansion joints will be used at each end of the pipe to provide sufficient flexibility to absorb the full pipe expansion. Two chemical feed systems and a blowdown tank have been included in the plant to provide a means of maintaining a total suspended solids concentration of less than 1,500 ppm as required by the American Boilermakers Association.

Fuel Oil Systems: The fuel oil systems consist of the oil system for the water-walled incinerator, and the oil system for the separately fired superheater. The separately fired superheater is to be located on General Electric property. This will require pipeline heating and a pump and heater set sized for approximately 15 gpm (gallons per minute). General Electric has sufficient oil storage capacity available to supply both this system and the system required for the water-walled incinerator.

Duplex pumps will be used to transfer the oil from General Electric's tanks to a 20,000 gallon tank located in Saugus. A bypass valve is provided downstream of these pumps to recirculate the oil back to General Electric's tanks when the day tank is full. Pipeline heating is provided for the entire length of pipe, including the portion crossing the river, to maintain an oil temperature of about 125°F. Beyond the day tank, a suction heater and a pump and heater set are provided to supply up to 50 gpm of oil to the boiler. This is sufficient oil to generate 400,000 pounds of steam per hour when burning oil alone.

Combustion-Air and Induced-Draft Systems: The combustion-air system is designed with three combustion-air fans: an underfire air fan, an overfire air fan and an oil burner fan. In addition, an induced-draft fan will transport the flue gas from the incinerator to the atmosphere. The overfire air fan will serve a dual function when associated with the spreader stoker, i.e., it will supply air to the air-swept nozzles, and also combustion air above the grates. In the reciprocating grate stoker arrangement, it will supply overfire air only.

The oil burner air fan will be directly connected to the oil burner and will modulate according to the burner's demands. The underfire air fan is to be automatically modulated to maintain a predetermined excess air. Since this is often a point of contention among manufacturers themselves, and between manufacturers and engineers, this point should be finalized in the design stage of the plant. The induced-draft fan will automatically maintain a slightly negative pressure in the water-walled incinerator to minimize heat loss through openings and doors.

Air Pollution Control Systems and Stacks: An electrostatic precipitator will be provided on the water-walled incinerator to reduce particulate emission below the level required by local, state and federal air pollution control codes. The electrostatic precipitator is a particularly desirable device since high efficiencies are possible. There is only a small pressure drop through the unit, and the particulate matter is collected dry.

On the separately fired superheater, a mechanical cyclone will be used. Since particulate emission from oil burning is relatively small, it has been found in the past that cyclone separators were sufficient to meet air pollution control codes. Due to the rapid changes in air pollution control codes, it is advisable that this decision be reviewed carefully at the design stage of this project to determine if a cyclone collector is adequate. If the cyclone separator is not acceptable, an electrostatic precipitator specifically designed for oil-fired boilers can be used. An electrostatic precipitator is much more expensive than a cyclone separator, however.

A double-walled stack of corrosion-resistant steel will be provided rather than an unlined stack. Because of the insulating effect brought about by the double-wall construction, the temperature of the gas will not drop to the dew point even at low load conditions. As a result, corrosion is appreciably reduced, and the life of the stack is extended considerably.

SPECIAL CONSIDERATIONS

Reciprocating Grate Stoker Versus Spreader Stoker: The reciprocating grate is sized on the basis of loading in pounds of refuse per hour per square foot of grate area. To avoid damage to the grate, a loading of 60 to 65 pounds per square foot per hour should be used when burning refuse. Foster Wheeler used a loading of 62 pounds of refuse per square foot per hour based on the horizontal projected area of the reciprocating grate stoker. Therefore, the reciprocating grate stoker considered is rated to burn 384 tons of refuse per day at the design loading.

In the unit incorporating the spreader stoker, much of the heat is released while the shredded material is in suspension. Therefore, more of the heat is absorbed by the boiler walls, and the grate is exposed to less heat than when a reciprocating grate stoker is used. As a result, the spreader stoker is capable of burning more refuse per day than a reciprocating grate stoker of the same physical size.

The spreader stoker is designed on a heat release of 750,000 Btu/hr.-ft.^{2} of grate area. Utilizing a standard size having 38% less horizontally projected grate area, the proposed spreader stoker can burn considerably more refuse than the reciprocating grate stoker. With present refuse having a heating value of 4,970 Btu/lb., the spreader stoker can burn 612 tpd as compared to only 384 tpd with the reciprocating stoker.

The spreader stoker is therefore evaluated for use in the proposed incineration plant, even though the refuse must be shredded prior to firing, since with relatively the same water-walled boiler, over 50% additional refuse can be burned with the spreader stoker. In addition, this means that less oil will be needed as supplementary fuel.

Dumping Floor Versus Storage Bin: The general ground elevation in the vicinity of the proposed process plant is less than 10 feet above mean tide level. Therefore, any subsurface construction would require expensive dewatering facilities. For this reason, it was considered to store the refuse on a flat slab rather than in a below ground storage bin. With the refuse on a flat slab, front-end loaders could be used to place the refuse onto the feed conveyors. Since front-end loaders are much less expensive than bridge cranes, this alternative was explored in depth.

Although expensive excavation could be eliminated by using the slab construction, the size of the storage area would be much larger, since it would not be practical to stockpile refuse higher than 10 feet with front-end loaders. Consequently, the building cost for the flat slab alternative would be only about $100,000 less than the building costs for the storage bin alternative. The bridge cranes are estimated to cost about $250,000 more than the front-end loaders. However, because the density of refuse on the floor will be less than in the pit, and because front-end loaders with a capacity of 8 cubic yards can handle only about one-half the amount of refuse a crane can handle in the same work period, the number of personnel required for the flat slab alternative is greater than the number required for the bin and crane alternative. The additional labor cost would amount to about $50,000 per year. It was estimated that the operation costs, other than labor, for the front-end loaders would be about $20,000 greater than the comparable operation costs for the cranes.

Based on an economic life of 30 years for the buildings, 20 years for the cranes, 5 years for the front-end loaders, and interest at 6%, $20,000 could be saved annually by using the bin and crane method rather than the flat slab and front-end loaders. From the viewpoints of public health and sanitation, confining the refuse to a below-grade pit would be much more desirable. For these reasons the bin and cranes will be used in all alternatives.

Shredding Versus Handling Raw Refuse: The shredded refuse system has several advantages over the nonshredded refuse system. When conveying the material, particularly over long distances, narrower belts, which are much less expensive both initially and from an operation and a maintenance standpoint, can be used with shredded refuse. The belts used in the nonshredded systems are

10 feet wide while those for the shredded refuse system are only 3 feet wide.

Shredded refuse can be stored in silos, and with a positive means of discharge, it can be metered to the incinerator. As a result, the feed to the incinerator is more uniform. Because of the smaller particle sizes, the burning process will be more complete, thereby improving the efficiency of the incineration process. A major factor in favor of shredding and utilizing a spreader stoker is the added incinerator capacity.

The primary disadvantage with total shredding is the added cost of owning and operating the shredders and the refuse storage silos. This cost is partially offset by the fact that the crane is operated for a shorter period of time per day, thereby reducing crane operation and labor costs. When burning raw refuse, there are no shredding costs and storage silo costs. Since, however, a spreader stoker cannot be used to burn unshredded refuse, the 612 tpd burning capacity cannot be achieved, and the plant would not be able to burn more than 384 tons of refuse per day with the proposed boiler and a reciprocating grate stoker. In addition, very wide belts must be used to convey the raw refuse.

COSTS

The value of the steam produced by the proposed boiler and superheater, based on General Electric's current production costs is $0.877 per 1,000 pounds. The annual cost for refuse disposal and steam generation would be in the vicinity of $2,500,000. Based on a value of $0.877 per 1,000 pounds for steam and an average General Electric demand of 200,000 pounds per hour, the steam would have an annual value of $1,480,000. For comparison with other methods of refuse disposal, the net cost of refuse disposal has been estimated to be $4.48 per ton.

COMMERCIAL DEVELOPMENT

A plant is expected to be built by RESCO, Inc. RESCO is a joint venture formed by Wheelabrator-Frye, Inc., and M. De Matteo Construction Co., of Quincy, Mass. The two partners expect to collect a total of 1,200 tons of trash daily from 16 surrounding communities. The plant, which should be operating by mid-1975 will turn this waste into as much as 350,000 lbs. of steam an hour. RESCO will supply steam at around 800°F. and 625 psi.

Wheelabrator-Frye is the U.S. licensee for Von Roll, Ltd., of Zurich, Switzerland, which has been building equipment for refuse burning power plants in Europe and Japan. Von Roll also has a Canadian licensee that that has built several garbage burning steam plants in Canada, including one in Montreal that is the same size as the plant Wheelabrator and De Matteo will build in Saugus.

Philadelphia, Pa.: Feasibility Study

This chapter is based upon a report prepared by the Envirogenics Company for the EPA in 1971, the purpose of which was to develop design recommendations and procedures for the disposal of refuse, a low sulfur fuel, with heat recovery in utility grade boilers.

WASTE MANAGEMENT OPERATIONS

The Sanitation Division of the City of Philadelphia's Department of Streets has cognizance of that City's waste management operations. According to its charter, the Sanitation Division is charged with six responsibilities, two of which are:

> "Collection of ashes, rubbish and garbage from households and retail establishments"

> "Disposal of all refuse removed by city forces by operation of incinerators and sanitary landfills. Also the disposal of combustible refuse collected by private contractors and industrial establishments and delivered to incinerators."

Fulfillment of these mandates is being accomplished by a collection system which functions within six sanitation areas, each of which is divided into two sanitation districts. Disposal is effected largely through the operation of six incinerators. The districting and incinerator locations are shown in Figure 4.1.

During the one year period from July 1, 1969 to June 30, 1970, the incinerator system fired 683,711 tons of refuse. Of this quantity, 12.4% was combustible refuse delivered by private operators and a small amount of special collection garbage brought in by city trucks. Another 200,245 tons of refuse was disposed of in landfill sites. It was estimated that about 20% of the landfill material was refuse of the oversized type.

Disposal costs, as of June 30, 1969, were estimated to be $7.35/ton by incineration and $1.74/ton by landfill. By definition, these costs do not of

FIGURE 4.1: PHILADELPHIA SANITATION AREAS AND INCINERATOR SYSTEM

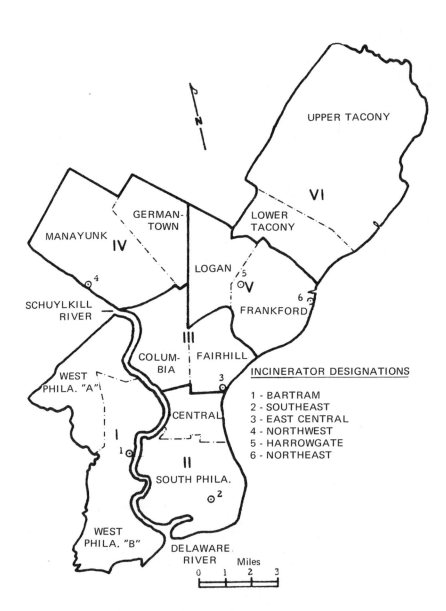

course include transportation, which in the case of landfill operations is substantial.

A third method of disposal used in Philadelphia is the outhaul of garbage to swine raisers. The demand, however, is not sufficient to absorb the entire production of garbage in Philadelphia. The districts in which separate garbage collections are made include: West Philadelphia "A", West Philadelphia "B" (excluding Eastwick), South Philadelphia, Manayunk, Germantown, Logan, Frankford, and Lower Tacony. The slop market has steadily declined with time due to the displacement of pig farming by rural urbanization. In the last decade, garbage deliveries to farmers dropped an estimated 42%. At the present rate of decline, the market could become nonexistent in another 12 years. Another factor which may make this operation even more short lived is that it requires subsidization by the City of Philadelphia. Elimination of the slop program has been frequently considered by city officials in recent years. It was therefore considered appropriate to assume that all garbage being collected separately would eventually be mixed with other refuse and be put into the regular disposal system.

Refuse Inventory and Composition Projections

Refuse Quantities: Collection and disposal statistics for Philadelphia are shown in Table 4.1 for the period July 1, 1969 to June 30, 1970. Included are all municipal operations involving either the collection or the disposal of solid wastes. Thus, data on solid wastes transported by private haulers for disposal in privately owned facilities is not included. This category is essentially of the commercial/industrial type. It will be noted, however, that some 8.2% of the total solid waste handled by the city is commercial/industrial. 90% of this material is fired in two of the city's six incinerators; these are the Bartram and the Southeastern Incinerators. The small amount of garbage brought in from special collections (in response to complaints, etc.) is also fired in the incinerators.

The amounts of refuse collected per unit area vary widely in Philadelphia. This can be seen from Table 4.2, which shows refuse densities by districts. Homogeneous distributions of the amounts of refuse collected within each district have been assumed. It is recognized of course that large nonresidential tracts exist, particularly in West Philadelphia B, South Philadelphia, and the Fairmont Park areas of West Philadelphia A and the Columbia District. The zone of greatest refuse output density is still obviously the block of five districts comprised of West Philadelphia A, Columbia, Fairhill, Germantown and Logan.

The quantities of refuse handled by municipal forces over the past decade and projected for the next decade are shown in Figure 4.2. This graph was prepared by the Sanitation Division of the Department of Streets and incorporated in an internal report of the City of Philadelphia. The data show the decrease of garbage collections for the slop market discussed earlier, and a decline in the amounts of commercial/industrial solid wastes being hauled to city incinerators. The latter effect does not mean that less industrial/commercial solid waste is being collected, but merely that private haulers are taking more of their collections to privately owned disposal sites.

TABLE 4.1: PHILADELPHIA REFUSE COLLECTION AND DISPOSAL STATISTICS BY DISTRICTS[1]

-------- Disposal Method, tpy --------

District/Source	Incinerators	City Owned Landfills	Contracted Landfills	Animal Feed[2,3]	Totals
West Phila. "A"	45,926	12,274	45,613	16,270	120,083
West Phila. "B"	51,169	0	0	7,040	58,209
Central City	27,756	6,424	0	0	34,180
South Phila.	78,983	22,025	5,122	0	106,130
Columbia	76,316	84	3,071	0	79,471
Fairhill	64,611	2,493	2,937	0	70,041
Logan	50,242	304	529	14,130	65,205
Germantown	56,396	1,939	1,239	16,310	75,884
Manayunk	42,754	1,755	942	12,440	57,891
Frankford	55,578	84	6,572	15,560	77,794
Upper Tacony	33,740	0	86,534	0	120,274
Lower Tacony	15,213	0	304	4,250	19,767
Spec. Garb. Coll'ns.	5,160	0	0	0	5,160
Comm'l./Ind.[2]	79,867	0	0	0	74,867
Totals	683,711	47,382	152,863	86,000	964,956
% of overall total	70.5	4.9	15.7	8.9	100

[1] 7/1/69 through 6/30/70. [2] Collected by private haulers.
[3] District garbage data are proportioned estimates. In West Phila. "B", allowance was made for Eastwick Station, where separate garbage collections are not made.

TABLE 4.2: PHILADELPHIA REFUSE DENSITIES BY DISTRICTS*

District	Total Solid Waste Collected, tpy	District Area, sq. mi.	Refuse Density, tpd/sq. mi.
West Phila. "A"	120,083	9.5	34.6
West Phila. "B"	58,209	15.4	10.4
Central City	34,180	4.4	21.3
South Phila.	106,130	13.1	22.2
Columbia	79,471	7.1	30.7
Fairhill	70,041	5.7	33.7
Logan	65,205	6.9	25.9
Germantown	75,884	7.1	29.3
Manayunk	57,891	15.6	10.2
Frankford	77,794	10.3	20.7
Upper Tacony	120,274	33.2	9.9
Lower Tacony	19,767	7.6	7.1
Entire city	884,929	135.9	17.8

*Based on collections for 7/1/69 through 6/30/70 and excluding commercial/indistrial and special garbage collections.

FIGURE 4.2: REFUSE COLLECTION RATES IN PHILADELPHIA

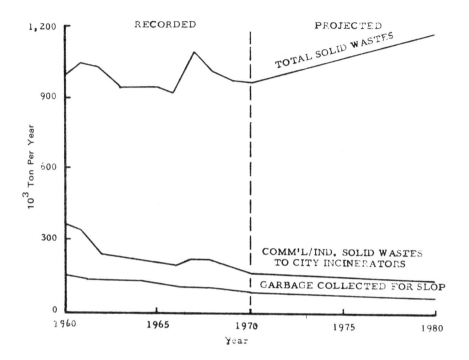

Similarly, the drop in garbage collection rates does not mean that the amount of garbage put out has decreased to that extent. It merely reflects the shrinking of the slop market and that separate garbage collections are giving way to mixed collections, the latter being disposed of in incinerators or landfills.

Figure 4.2 also shows an erratic trend in total waste disposal, which actually reflects a slight net drop over the ten year period. This is because the rate of decline in commercial/industrial receipts at the incinerators has been so much greater than the rate of increase in other urban collections (by far the larger fraction) as to create a small decline in the overall disposal rate. If the industrial/commercial fraction is extracted from these data, an average increase of 1.39%/year for other urban refuse collections is found.

This is considerably lower than the 3.0%/year predicted for the nation on the initial Envirogenics' program. However, this rate involves two factors: (1) a per capita refuse production increase of 1.5%/year, and (2) a 1.5% increase in the national population. The population in Philadelphia decreased during the period 1960-1970 from 2,002,512 to 1,948,609 according to Bureau of Census figures. This represents an annual population drop of 0.27%. If this negative rate is combined with the expected rate of increase in the refuse collected per capita (1.5%/year), the resultant estimated increase in domestic refuse collections for

1970 would be 1.23%. This compares reasonably well with the 1.39% actually observed.

It will be noted in Figure 4.2, that the changes in collection rates projected by Philadelphia's Sanitation Division for the present decade do not appear to be consistent with those of the previous decade.

Projected and Actual Philadelphia Refuse Handling Rate Changes

	Decade of the Sixties (Recorded)	Decade of the Seventies (Projected)
Change in Urban refuse collections,* %/year	1.39	2.06
Change in commercial/industrial solid waste receipts at city incinerators, %/year	−9.56	−0.69
Change in garbage collections for farm consumers, %/year	−5.09	−2.00
Change in total quantities of solid waste handled by city forces and/or facilities and city contracts, %/year	−0.30	2.00

*Including garbage for farm use but excluding commercial/industrial.

What is actually involved in quite plausible. During the previous decade a sharp decline in commercial/industrial solid waste receipts cancelled the normal increase in other urban refuse collections so that the total waste handled remained about even. In the present decade, however, the Sanitation Division policies regarding incinerator access by private haulers are expected to level out the receipt rates of commercial/industrial waste material. This will then result in an upward trend in the curve of total solid waste collected, which will reflect the increase in other urban collections. It is questionable, however, whether the urban collection rate would increase from 1.39 to 2.06%/year.

In the preceding discussion, the rate of change in the quantity of garbage collected for farm use was not considered. This is because this operation has no influence on the other rates. The total solid waste collection rate and the rate at which urban refuse is collected involve all the garbage that is hauled. These rates are therefore insensitive to the manner in which the garbage is ultimately disposed of or recycled. In terms of firing refuse in steam generators, however, this factor must be taken into account. Obviously, any garbage taken to farm users must be excluded in considering the quantities of refuse that will be available for firing in boilers.

A twenty-year projection of the quantities of refuse fuel that will be available in Philadelphia from city managed sources is shown in Figure 4.3. The

FIGURE 4.3: PROJECTED PHILADELPHIA BOILER REFUSE-FUEL INVENTORIES

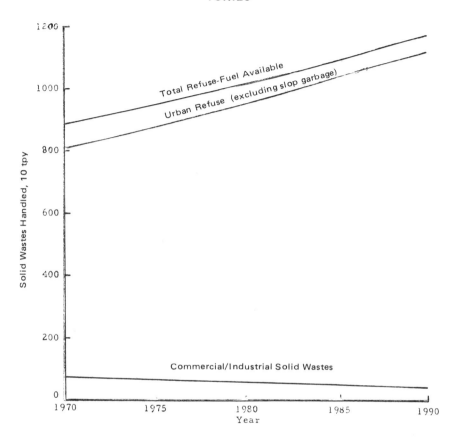

projections for commercial/industrial solid waste receipts and garbage hauled to farms made by the Sanitation Division have been observed. The rate of increase in urban collections has, however, been based on that (1.39%/year) recorded for the previous decade. It should be mentioned that the amount of slop garbage collected by 1980 will have dropped to about 50,000 tpy. This is probably approaching the level where the practice could well be discontinued as economically unattractive.

Refuse Composition: Establishment of the composition of Philadelphia fuel-refuse is complicated by the fact that two types exist in the various districts. That is, urban refuse with and without garbage is being collected. Because the duration of this situation is uncertain and because the two types would be difficult to isolate in disposal operations, the use of composite data appeared to be most practical.

Figure 4.4, showing compositional projections for the period 1970-1990, has been prepared on that basis. The data were obtained using the compositional

FIGURE 4.4: PROJECTED COMPOSITIONAL CHANGES IN PHILADELPHIA MIXED REFUSE

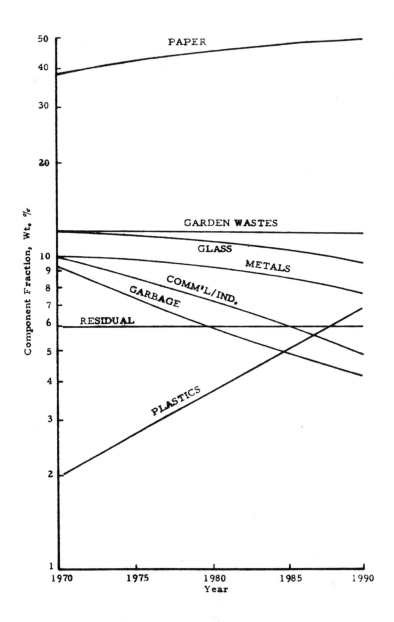

change derivations developed on the original Envirogenics program and the Philadelphia disposal trends previously discussed.

Fuel Characteristics of Philadelphia Refuse

Heating Value: Using the compositional values discussed in the previous section and the component calorific values adopted on the previous Envirogenics program, higher heating values (HHV) were calculated for mixed Philadelphia refuse. These are shown in Figure 4.5 for the period 1970-1990. An increase in the projected HHV for refuse results because the predicted compositional changes (Figure 4.4) involve increases in the levels of high HHV constituents, paper and plastics, and decreases in low HHV components, such as glass, metals, and garbage.

Work is being done at the Drexel Institute to determine the calorific value of refuse fired at certain of the City's incinerators. In a private communication, Dr. R. Schoenberger of that Institute advised that a current typical value would be about 5500 Btu/lb. The material tested was, however, garbage-free.

Combustion Calculations: In developing preliminary designs of combination-fired systems, some specific refuse composition must be assumed. It is recognized that the design fuel would not be valid except for a comparatively short period of time. It would, however, furnish a reference point for making adjustments in firing rates, etc., when fuel characteristics are significantly different. For the purposes of this study, the refuse composition projected for 1980 was used. It is tabulated below for easy reference.

Projected 1980 Composition (Wt.-%) of Mixed Philadelphia Refuse

Garbage	Plastic	Garden Wastes	Glass	Metals	Paper	Residual	Commercial/Industrial
5.9	3.7	12.0	11.2	9.2	44.8	6.0	7.2

The heating value, as shown in Figure 4.5, will be about 4,700 Btu/lb. Using the same guidelines as observed on the previous program, the following ultimate analysis was computed.

Projected 1980 Ultimate Analysis of Mixed Philadelphia Refuse

Component	Weight-Percent
H_2O	17.2
C	26.8
H	3.5
O	22.9
N	0.4
S	0.2
Inert	29.0
	100.0

Combustion air, flue gas, and steam generator efficiency were then calculated. The results are shown in Tables 4.3, 4.4 and 4.5. It will be noted in Table 4.5 that steam generator efficiencies have been calculated on the basis of three different flue gas exit-temperatures.

FIGURE 4.5: PROJECTED INCREASE IN HEATING VALUE OF MIXED PHILADELPHIA REFUSE (AS RECEIVED BASIS)

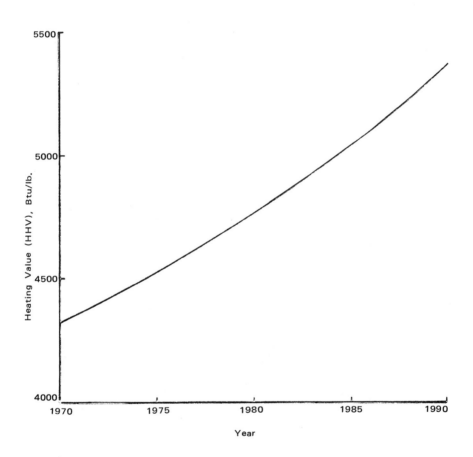

TABLE 4.3: COMBUSTION GAS REQUIREMENTS BASED ON PROJECTED 1980 PHILADELPHIA MIXED REFUSE COMPOSITION

Constituent	Combustion Gas Requirement, lb./lb. Refuse	
	Oxygen	Dry Air
C	0.714	3.077
H	0.280	1.207
S	0.002	0.009
Metal	0.013	0.056
	1.009	4.349
Oxygen	−0.229	−0.987
Total required, stoichiometric	0.780	3.362
Total required, 50% excess gas	1.170	5.043
Excess gas	0.390	1.681

TABLE 4.4: PRODUCTS OF REFUSE COMBUSTION BASED ON PROJECTED 1980 PHILADELPHIA MIXED REFUSE COMPOSITION

Constituent	Gas Formed per lb. Refuse		Volume Percent Dry Basis
	Lb. Mol	Lb.	
CO_2	0.022	0.983	12.8
H_2O	(0.030)	(0.544)	–
Refuse H_2	0.018	0.315	–
Refuse H_2O	0.010	0.172	–
Combustion air*	0.003	0.057	–
SO_2	0.001	0.004	0.03
O_2 (excess)	0.012	0.390	7.0
N_2			80.23
Total air N_2	0.138	3.876	
Refuse N_2	0.001	0.004	
Total flue gas (wet)	0.202	5.801	
Total flue gas (dry)	0.172	5.257	

*Based on standard (60% RH @ 90°F.) of American Boilers Manufacturers Association.

TABLE 4.5: EFFICIENCY OF STEAM GENERATOR FIRING PHILADELPHIA MIXED REFUSE OF COMPOSITION PROJECTED FOR 1980

Item	Fuel Value Heat Losses at Various Flue Gas Exit Temperatures, Btu/lb. of Fuel (% of HHV)		
	450°F.	500°F.	575°F.
Dry gas	467 (9.79)	530 (11.11)	625 (13.10)
H_2O in refuse	209 (4.38)	213 (4.46)	219 (4.59)
From H_2 combustion	383 (8.03)	390 (8.18)	402 (8.43)
H_2O in air	12 (0.25)	13 (0.27)	16 (0.34)
Unburned gas	4 (0.08)	4 (0.08)	4 (0.08)
Unburned residue	107 (2.24)	107 (2.24)	107 (2.24)
Sensible heat, residue	47 (0.98)	47 (0.98)	47 (0.98)
Unburned fly ash	40 (0.84)	40 (0.84)	40 (0.84)
Sensible heat in fly ash	5 (0.10)	5 (0.10)	6 (0.13)
Subtotal	1,274 (26.69)	1,349 (28.26)	1,466 (30.73)
Radiation	(0.20)	(0.20)	(0.20)
Unmeasured	(0.50)	(0.50)	(0.50)
Manufacturer's margin	(1.00)	(1.00)	(1.00)
Total % heat loss	28.39	29.96	32.43
Steam gen. efficiency	71.61	70.04	67.57

Note: Fuel Value (HHV) = 4770 Btu/lb.; Combustion Air Inlet Temperature = 80°F. (60% RH).

Utility Steam Generation Operations

Steam Generator Inventory: The Philadelphia Electric Co. (PE) provides electrical power and some heating steam for the City. The PE inventory of power stations is widely distributed. Within the city limits, however, there are four power stations and two steam plants. One of the units at the Schuylkill Plant is equipped with a topping turbine. It is therefore linked with the two downtown steam plants in the Center City heating steam loop. This circuitry is discussed in a later section.

The effective electrical capacity of the power stations located within the city limits is about a third that of the entire PE system, which includes nine other stations, all well outside the city limits. The Peach Bottom atomic station, for example, is almost 60 miles from downtown Philadelphia. The basic characteristics of the four power stations located within the city limits are summarized in Table 4.6. The characteristics of the two steam plants, Willow and Edison, are shown in Table 4.7. Units 23 and 24 of the Schuylkill Station appear in both tables. This is because they are coupled to a topping turbine and thus produce both electricity and heating steam.

The quantity of steam sent out from these units can be controlled by using a low pressure 20 MW turbine (throttle condition, 200 psig, 440°F.) in tandem with the topping turbine. Thus, any steam not required for the city heating system can be diverted into this second turbine to produce electricity. Over the past five years, the Schuylkill has accounted for an average of 64.4% of the district heating steam sent out. All three plants divert about 15% of their production to heat feed water and drive auxiliary plant machinery.

TABLE 4.6: UTILITY POWER BOILER INVENTORY WITHIN THE CITY OF PHILADELPHIA

Station (Effective Capacity*, MW)	Boiler No.	Pressure, psig	Temperature, °F.	Fuel
Southwark (462)	11, 12	925	900	Pulv. coal/oil
	21, 22	925	900	Pulv. coal/oil
Schuylkill (335)	1	2,475	1,050	Oil
	2, 3	225	545	Oil**
	11-20	225	503	Oil**
	23-24	1,350	910	Pulv. coal/oil
Delaware (422)	13-24	267	637	Oil**
	71-81	1,875	1,000	Pulv. coal/oil
Richmond (464)	49-52	400	703	Oil**
	57-60	400	703	Oil**
	63, 64	1,335	950	Pulv. coal/oil
	65, 66	425	850	Pulv. coal/oil

*Effective capacity for 75% of the year.
**Converted from coal firing.

TABLE 4.7: UTILITY STEAM PLANT INVENTORY WITHIN THE CITY OF PHILADELPHIA

Station	Boiler No.	Rated Steam Prod., 10^3 lb./hr.	Pressure, psig	Temperature, °F.	Fuel
Edison	1, 2	216 ea.	205	435	Oil
Willow	1-3	125 ea.	200	438	Oil*
	4	170	180	434	Oil*
	5	170	190	430	Oil*
	6	170	180	434	Oil*
Schuylkill	23, 24	600 ea.**	225	450	Pulv. coal/oil

*Converted from coal firing.
**Output of topping turbine.

Effect of Firing Refuse on Pollution Burden: Because of its predominant use in utility-class boilers in Philadelphia, low sulfur (approximately 0.5% S) oil would likely be the fuel that would in effect be substituted for if refuse were also fired in boilers. The particulates and SO_2 pollutants emitted by refuse would be about a fifth and a half, respectively, that emitted by low sulfur oil, on an equivalent energy basis. Assuming, for refuse and oil, fuel values of 4,770 and 15,000 Btu/lb. and boiler efficiencies of 70 and 88%, respectively, this represents an available energy ratio of about one to four. It also assumes that no flue gas cleaning would be practiced when oil is fired but that a refuse-fired steam generator would incorporate an electrostatic precipitator having an efficiency of 99%.

A ton of fired refuse will produce, on the average, 2.3 lb. SO_2 while a quarter ton of oil (0.5% S), the equivalent in available energy, will produce about 4.9 lbs. of SO_2. Thus, if all the refuse presently collected by city forces (875,000 tpy) were substituted for low sulfur oil, the annual output of SO_2 would be reduced by slightly over 1,000 tons.

In terms of particulates, refuse fired on agitating grates loads the flue gas with about 1.3 gr./scfd of fly ash, while oil produces a loading of only about 0.1 gr./scfd. Differences in excess air requirements considered, oil produces 2 1/2 times as much flue gas as does refuse. This, taken together with a refuse:oil available energy ratio of 1:4, suggests that refuse will produce 20 times as much fly ash as does oil in developing the same amount of steam enthalpy.

Typically, however, a refuse-fired system will be equipped with gas cleaning equipment while an oil fired boiler would not. In the former case, an electrostatic precipitator having an efficiency of 99% would be reasonable to expect. In this situation, the refuse-fired system would produce about 20% the particulates of an uncontrolled oil-fired boiler producing the same amount of steam energy. This would represent a reduction of 0.96 lb. of particulates for every ton of refuse fired in replacement of oil, an air pollution burden relief of about 420 tons per year, based on present refuse collection rates (875,000 tpy).

It should be pointed out that, in the foregoing discussion, no account has been taken of the fact that considerable refuse is presently being fired in conventional incinerators in Philadelphia. Although the overall dust control efficiency of this system of incinerators is not known, it is said to be considerably less than 99%. An additional benefit would thus result from firing the same refuse in steam generators equipped with high efficiency gas cleaners.

PRELIMINARY PLANNING RECOMMENDATIONS

Overview

In developing the following systems recommendations, it was necessary to adopt and follow certain general guidelines. These are itemized below.

Lead Time: The typical period of time elapsed between construction go-ahead and initial service of a conventional power plant is about seven years. Because of the less conventional nature of refuse-firing systems, 1980 was set as a convenient target date. The refuse characteristics and quantities projected for that time have been discussed earlier.

System Input: The overall system recommended should be capable of handling all of the refuse for which the city will be responsible in 1980 and preferably allow for expanded throughput beyond that date. The system should, however, comprise more than one plant so that initial capital cost burdens can be spaced out and so that logistics are manageable. In order that collection forces will have a reliable disposal operation to accept their deliveries, a high plant factor will be necessary. This has been set at 80%, as on the previous Envirogenics' program. During outages, elements of the existing incinerator/-

landfill disposal system would be substituted.

Plant Management: In terms of PE and City participation, many management options can be considered. Attempting to influence the decisions involving such alternatives is clearly not an objective of the program. For costing purposes, however, the capital cost annualization rate selected was that associated with utility ownership. This was done merely in the interest of conservatism. The average utility annualization rate is usually considerably higher than would be obtained under municipal ownership.

District Steam Plant

Design Characteristics: The plants Edison, Willow, and Schuylkill pump steam into an essentially common distribution loop, the highest steam condition input being from Schuylkill at 225 psig and 450°F. (enthalpy approximately 1235 Btu/lb.). The lowest steam demand is of course during the summer months when the steam send-out is below 400,000 lb./hr., little if any of which goes into space heating systems.

The Philadelphia Electric Co. has demonstrated that it is more economical to operate at 100% feedwater make-up than to attempt recycle of the condensate, which becomes heavily contaminated by the district heating circuitry. Because the boilers operate at low steam temperatures, feedwater treatment can be limited to a softening process rather than deionization, although the boiler must then be blown down fairly frequently. A final factor influencing design is that Philadelphia's district heating requirement will probably not increase significantly with time, because the central city served will not be involved in much further growth.

Thus, a refuse-firing, district-heating plant can be designed on the basis of today's needs. If sized to summer demand, such a plant can be operated at base-load condition all year around, assuming that, of the three existing plants, only the Willow plant would be fired during the summer months to fill the northern segment of the steam loop and thus reduce the frictional line losses that would result from single plant input. A send-out of 130,000 lb./hour would be sufficient to accomplish this.

It was suggested by the PE that the refuse-fired plant have a steam production of no more than 350,000 lb./hour. This is equivalent to a send-out of about 300,000 lb./hour. This size, however, involves a refuse input that would probably require the use of three boilers. By dropping the capacity to about 315,000 lb./hour, two boilers would be sufficient. Production costs would thus be significantly reduced. In order to generate steam at a rate of 315,000 lb./hour (send-out approximately 270,000 lb./hour), a refuse rate of 1,400 tpd would be required. Fuel characteristics would, as stated earlier, be based on those projected for 1980.

The plant would consist of a completely indoor structure, housing two units in a side-by-side arrangement. Both units would face upon a common storage pit, which would be separated from the boilers by a 50-ft. fire wall running the full length of the pit.

Overhead cranes would transfer refuse from the pit to a water-cooled charging chute on each of the units. Stoking would be promoted by means of a ram or vibratory feeder. In each furnace the grate would be a three-stage, reciprocating device, the ash from which would be quenched within the ash pit by water sprays. Combustion air, 50% in excess of stoichiometric, would be drawn from the refuse pit area so that odor leakage would be minimized. From 60 to 75% of the total air introduced into each unit would be directed to the underfire and sidefire jets. This air would first be heated by a tubular air heater to about 325°F. Overfire air would be introduced so that the hot combustion gases would flow back over the bed and tend to dehydrate the material on the first grate stage. Natural gas burners would be provided, but for start-up and trimming purposes only.

Because of the heterogeneity of the fuel, steam conditions would be more variable than those experienced in firing fossil fuel. In the case of turbo-electric plants, such fluctuations are unacceptable so that other design arrangements are necessary. However, based upon European and domestic experience, this variability is regarded as acceptable for district heating steam and no provisions are needed for fuel augmentation.

In the radiative section of each boiler, standard water wall construction would be employed. This would extend below the grate and partially under it. Because of steam use and radiative super-heating, the boilers would have, respectively, no pendant reheat or super-heat surfaces. Most of the convective section would be occupied by the economizer and the tube banks communicating with the mud drum. The temperature of the flue gas exiting from the air heater would be about 500°F. Thus, the furnace efficiency (see Table 4.5) would be about 70.0%. The flue gas from each unit would be cleaned in a separate electrostatic precipitator having a dust removal efficiency of 99% and each sized to handle an expected flue gas flow of 135,000 acfm. After cleaning, the two flows would be blended and released from a common stack. General plant lay-out is shown in Figure 4.6.

Because of the low sulfur content of the refuse fuel (approximately 0.1% S) and the fact that fossil fuel would not ordinarily be fired in the furnaces, provision for an SO_2 control system was considered unnecessary. The plant could be expected to emit about 2,800 lbs./day of SO_2 and about 350 lbs./day of particulates. This would correspond to a stack gas composition of 0.012 gr./scf and 0.008 vol.% with respect to particulates and SO_2.

On this basis, it is assumed that refuse would generate 24 lb. of furnace fly ash per ton of refuse fired and that the sulfur content of refuse would be 0.1% S, only half of which would be converted to SO_2. In view of past experience, the expectation of achieving a 99% collection efficiency appears to be reasonable. However, considerable tolerance is available if one compares the expected dust output of 0.24 lb./ton of refuse with the projected national particulate emission standard of 1.9 lb./ton of refuse fired.

Feedwater would be introduced at 220°F. This would be heated in open-type, deaerating feedwater heaters equipped with vent condensers. Steam from the boilers would be used as the heat source. To effect pressure reduction, the steam would first be used to drive the turbines on the boiler feed pumps. It is assumed that 100% feedwater make-up would be practiced and that the water would be softened (through synthetic zeolite) rather than deionized. Salt accumulations within each boiler would be controlled by frequent blow-downs of the oversized mud drum.

FIGURE 4.6: LAYOUT OF STEAM PLANT

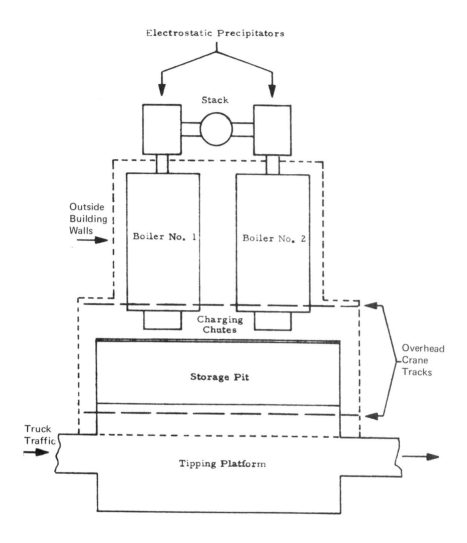

The tipping pit would be sized to permit continuous 7-day-a-week firing. Assuming that all packer-truck deliveries are accomplished between 10 A.M. Monday and 12 noon Friday, and allowing an additional margin of four hours firing time, the pit should be able to accommodate 4,300 tons or 6.60×10^5 cu. ft. of refuse. This would require a pit dimension of 200(l.) x 55(w.) x 60(d.) cu. ft.

Three 100-ft. bridge cranes would be installed over the pit, one of which would be used to arrange and mix the pit contents during the receiving hours (day shifts, Monday through Friday). During other shifts, it would be held in stand-by without an operator. Assuming a charging cycle of four minutes, grapples having a capacity of 5 cu. yd. would be sufficient. The lift speed of the cranes should be at least 300 ft./min. and provide a horizontal travel speed of at least 350 ft./min.

The storage pit would be able to accommodate 13 tipping stations. Assuming, conservatively, an average load of 4 tons (15 cu. yds.) per truck and a discharge cycle time of four minutes, an off-load rate of 780 tph would be reasonable to expect. The maximum receiving rate should, however, not exceed 520 tph. This is based on the assumption that all the refuse fired would be delivered according to a 5-day collection schedule (1,900 tpd) and that 80% of the daily receipts would be delivered during two 1.5 hour peak periods. Thus, no truck queuing is likely. Weigh-in would be handled by a single automated scale; weigh-out, which is required by the City of Philadelphia, would be accomplished on a second identical scale.

Disposal of bulky refuse should probably also be handled at this plant. This is because it will probably have a centralized location due to its association with the mid-city steam loop. The smallest shredder capable of accepting typical bulky refuse items would have a rating of about 700 hp. Such a machine could coarsely (7-10 in. top size) reduce bulky items at a rate of at least 30 tph. Thus a single shredder would probably handle the entire city's output of such refuse, which will likely amount to some 250 tpd by 1980.

Costs: The cost model developed on the initial program for the EPA was modified extensively. This was done to accommodate cost increases that have occurred since the development of the model and economic factors characteristics of the Philadelphia area. It was also necessary to introduce or substitute new cost elements. These involved design features common to many steam plants (e.g., feed water treatment and heating equipment) that are different from those found in turbo-electric systems. Another costing change arose from the fact that refuse-charging equipment was based on the use of cranes rather than live-bottom storage structures and conveyors, as was done on the original program. The highlights of these changes, as recommended by City and PE officials, are as follows:

(1) *Land Cost:* This was increased from $10,000 to $40,000 per acre, based on the sites considered.
(2) *Federal Power Commission (FPC) Boiler Component Codes 311-316 Costs:* These and certain other specified capital costs were increased at a rate of 10% per year based on the cost model development date, June 1969. The new base date has therefore been shifted to June 1971.
(3) *Annualization Rate:* This was reduced to 13.75%.
(4) *Water Cost:* This was set at $4.5¢/10^3$ lb. based on steam send-out.
(5) *Maintenance Costs:* The present PE cost is $12¢/10^3$ lb. of steam produced. To this was added another 25% in consideration of maintenance problems unique to refuse-firing and the separate costs for maintaining shredders and APC equipment.

(6) *Refuse Storage Pit Costs:* These were developed on the basis that the pit should be structurally stable if completely flooded with water.

(7) *Residue Disposal Costs:* This was based on the Philadelphia practice of using truck outhaul. In the present plant this would involve four drivers and six trucks (two standing under hoppers).

(8) *Fuel Costs:* Because the price of fuel oil is unstable at this time, refuse disposal costs have been presented as a function of this cost over the range $0.30 to $0.60 per 10^6 Btu's.

Costs for the district heating plant are shown in Table 4.8. Total refuse disposal costs is the difference between total annual costs and the annual credit for steam generated. Unit refuse disposal cost is this difference divided by the quantity of refuse handled each year, which is 0.409×10^6 tons, assuming an 80% plant factor. This cost will vary considerably, depending on how the annual credit for steam is derived.

TABLE 4.8: COSTS FOR THE 1,400 TPD PHILADELPHIA DISTRICT HEATING PLANT

FPC Codes	Description	Cost, 10^6 $
	Capital Costs	
310	Land and land rights	1.696
311	Structures and improvements	1.090
312	Boiler plant equipment	6.656
315	Accessory electrical equipment	0.531
316	Misc. power plant equipment	0.131
	Air pollution control equipment (98% efficiency)	0.486
	Waste handling equipment	2.105
	Engineering and inspection	1.174
	Total capital cost	13.869
	Annual Costs	
	Annual capital cost	1.907
	(Effective annualization rate = 13.75%)	
	Water cost	0.085
	Operating labor	0.578
	Maintenance	0.351
	Residue disposal	0.300
	Total annual costs	3.221

If the new refuse-firing steam plant is added to an existing inventory that is already capable of handling the demand, then the steam credit assignable to the new boiler can only reflect the cost of fossil fuel saved and the O&M costs transferred from the now less active, conventional boilers. If, on the other hand, the refuse-fired plant is added to the system to supply needed additional capacity or to permit the retirement of old equipment, the steam credit should reflect annualized capital costs as well. It was the judgment of the PE that the former situation prevails and that the steam production costs of existing plants, preferably the Schuylkill plant, be used to calculate refuse disposal cost.

As can be seen in Figure 4.7, the steam production cost at Schuylkill was set at $0.55/10^3$ lb. This would result in a refuse disposal cost of $5.36/ton, excluding transportation. As stated earlier, the steam sent out from Schuylkill issues from a topping turbine. Because of this arrangement, all plant labor costs are applied to power production.

Thus the Schuylkill steam production cost includes only a fuel cost equivalent to the energy content of the output steam, some supervision, and a small amount of maintenance. As can be seen from Figure 4.7, if production costs (first ten months of 1970) of PE's straight steam plants are used, considerably lower refuse disposal costs result. Also shown, for comparative purposes, is the tariff steam rate or the official rate of charge approved by the PUC. The rate selected (excluding state tax) is that for large Rate "S" steam users during the minimum demand period, June through September.

FIGURE 4.7: DISTRICT HEATING PLANT – REFUSE DISPOSAL COSTS AS A FUNCTION OF CREDITABLE STEAM VALUE

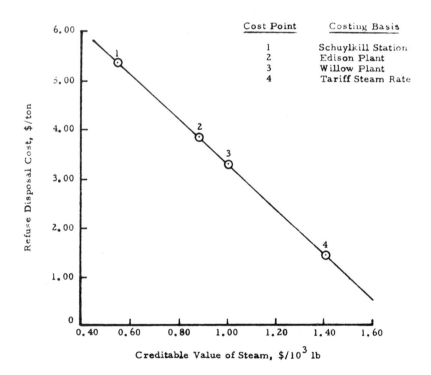

Philadelphia, Pa.: Feasibility Study

Site Selection: Because of the built-up nature of Philadelphia, it will be difficult to find suitable tracts of land that are reasonably close to the areas where the production of refuse is the greatest. An obvious solution to this problem would be to locate the steam generator plants at the sites where refuse incinerators are now located. With the commissioning of the refuse-firing steam generators, some of the incinerators would have little function except, perhaps, during outages of refuse-fired boilers. Thus the razing of one or two of the incinerators should be acceptable, even if interim refuse disposal by landfilling is required during the construction period. Another possibility is the use of the tract of land occupied by the Schuylkill Arsenal. This facility is adjacent to the southern property line of the Schuylkill Station and its purchase is being actively pursued by the PE.

In the case of the district heating plant, the site selected should obviously be located close to the existing steam lines used by the Philadelphia Electric Co. The scope of that system, excluding the feeder lines, can be seen in Figure 4.8. Considering first the incinerator sites, it will be noted that the East Central incinerator is somewhat closer to the loop (i.e., The Willow Steam Station) than the Bartram incinerator is to the loop on the east side of the Schuylkill River.

FIGURE 4.8: STEAM DISTRIBUTION SYSTEM OF THE PHILADELPHIA ELECTRIC COMPANY

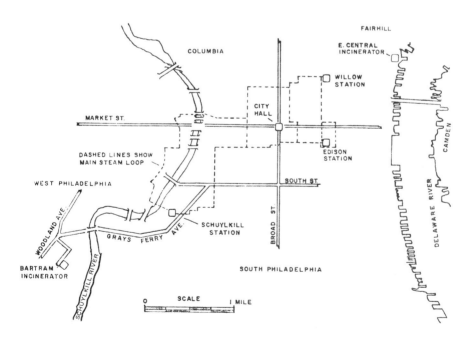

The East Central plant is separated from the loop, however, by a heavily built-up section of center city. Leading a steam line from it to the Willow Steam Station would be a major undertaking. The Bartram plant site provides a more practical easement to the loop. The third potential site, being continguous with the Schuylkill Station, is not shown. It is obviously, however, the preferred site in terms of steam loop access. Its principal drawbacks are that: (1) unlike the city incinerator sites, the land would have to be purchased; and (2) other uses for this plot are being considered by the PE. It was their advice, however, that the arsenal grounds be given primary attention on this program.

Transportation Costs: As shown in Figure 4.1, each sanitation area in the city is divided into two districts, many of which are designated by the name of a well-known section of town located within them. The amount of refuse hauled out of each of the districts to both landfill and the incinerators was tabulated in an earlier section. From these data, it was possible to estimate comparative transportation costs for hauling refuse, once the trucks are filled on their collection routes, to the existing incinerators and landfill sites, and to the Schuylkill site discussed in the preceding section. The assumptions adopted were that the refuse production densities within each of the districts were uniformly distributed therein and that landfill hauls averaged 10 miles per round trip. The latter value is a rough approximation.

Each of the Philadelphia incinerators fires refuse from several districts. None, for example, receives from fewer than seven of the twelve districts, while the Northeast Incinerator receives from nine. Each district was therefore roughly divided into zones, the area and location of which were selected to correspond with the logical direction of travel to and the proportion of refuse sent out to the various incinerators. The approximate center points of these zones thus served as loci for deriving weight-distance vectors for each district.

The travel distance on a grid makeup of surface streets would typically be the sum of the two orthogonal sides of a right triangle. This sum can be anywhere from 0 to 41% larger than the direct or diagonal distance. As a first approximation, the diagonal distance was increased by 25%. Distance travelled included return trip mileage. Thus in multiplying distances by tonnage the resultant ton-miles is actually about twice the real work performed on the refuse.

From these calculations it was estimated that for the period July 1, 1969 to June 30, 1970 the refuse transportation performed in Philadelphia was 5.8×10^6 ton-miles. Assuming a cost of \$0.25/ton-mile, the transportation cost was found to be \$1.81/ton.

A similar analysis was performed for the proposed district heating plant. This was done on the basis of the same haul cost (\$0.25/ton-mile) used above so as to permit direct comparison to be made. The selection of the sanitation districts to be served by the steam plant was based on refuse collection rates projected for 1980 and the proposition that no refuse produced in these districts would be disposed of by landfill. The plant would be adjacent to the Schuylkill Station. Fortuitously, it was found that the six districts in the central and southern portions of the city would provide just slightly more than the 1,400 tpd required to operate the plant. The refuse transportation data derived for these six districts is tabulated below.

Steam Plant Refuse Transportation Data

District	Refuse Hauled, ton-miles/year
West Philadelphia "A"	1,275,200
West Philadelphia "B"	394,000
South Philadelphia	670,100
Central	171,700
Columbia	844,000
Fairhill	814,100
Total	4,169,100

The projected total tonnage would be 537,400 tons/yr. The transportation cost would thus be $1.93/ton, which compares favorably with the $1.81/ton estimated for the present incineration system.

Power Plant

Design Characteristics: The steam plant described in the previous sections would be capable of handling about one-half of the refuse that will probably be collected in Philadelphia in 1980. It has been estimated that the refuse available for boiler fuel (see Figure 4.2) at that time will be about 1.02×10^6 tpy, which is equivalent to slightly over 2,800 tpd. It would therefore be appropriate to consider a combined-fired power plant that would have a refuse capacity either the same as or perhaps greater than the steam plant. If oversized, fossil fuel could be substituted for the refuse that would be lacking until such time as the growing collection rates could satisfy design input.

TABLE 4.9: CHARACTERISTICS OF 300 MW, CASE 3 POWER SYSTEM

Item	
Refuse rate (fractional heat input), %	19.4
Steam pressure, psig	2,400
Number of turbines	1
Total turbine heat input, 10^9 Btu/hr.	2.52
Steam generator efficiency due to refuse, %	67.6
Steam generator efficiency due to fossil fuel, %	87.0
Net steam generator efficiency, %	83.2
Heat input total, 10^9 Btu/hr.	3.036
Heat input from refuse, 10^9 Btu/hr.	0.596
Heat input from fossil fuel, 10^9 Btu/hr.	2.440
Refuse rate (firing 2 economizers), tpd	1,500
Fossil fuel rate (as coal), tpd	2,440
Net plant heat rate, Btu/kwh	10,120

Using the designations developed on the original program, a Case 3 system equipped with two refuse-fired economizers would require about 1,500 tpd of refuse. This is based on the heating value and boiler efficiency for the fuel (at an exit temperature of 575°F.) projected for 1980. The Case 3 system consists of a conventionally fired steam generator having little economizer surface and one or more externally situated "boilers." The latter are fired with refuse to deliver high enthalpy feedwater to the drum of the steam generator. This design was found on the previous program to be optimum in terms of cost effectiveness for the range of refuse rates in which the present one falls. A summary of the boiler characteristics is shown in Table 4.9.

Except for a slightly larger refuse pit, the refuse handling and charging arrangement for the two economizers in the system would be identical to that described for the district heating plant.

The two identical economizers would be operated in parallel and thus perform the same function. In each, refuse would be fed through a vertical, water-cooled chute and burn on a thick fuel bed. The three-level grates incorporated should furnish both agitation and tumbling to the fuel mass to insure good burnout. High velocity, secondary air nozzles would be provided in the front and rear walls to promote complete combustion of volatile gases and particles rising from the fuel bed. All walls and the roof would be of welded tube-and-fin construction.

Tube banks, especially in areas of relatively high gas temperatures, would be arrayed vertically. Horizontal tube banks would be of bare tube design in all cases. A tubular air heater, in which the flue gas would be directed downward inside the tubes, would be used because of its ease of cleaning. Ash hoppers would be appropriately located to remove ash where tube banks might act as ash deflectors.

Feedwater at 470°F. would be flowed in a single continuous (once through) path. Flue gas would be directed in a two-pass arrangement and be discharged into a dust collector located at grade level. Air, preheated to 316°F. would be delivered as underfire air. This temperature was selected as being compatible with the cast iron grate. Approximately 25% of the preheated air would be sent through a booster fan and delivered as high velocity secondary air. A 50% excess of air would be employed and the exit flue gas temperature would be 575°F.

The water wall panels would consist of 3-in. OD tubes spaced on 3-3/4 in. centers with fins continuously welded between tubes. Because of the all-metal construction, slag adhesion should be minimal. Gas-borne, molten slag-particles would be cooled upon contacting the tube or fin and thus tend to shed from the surface. The solid walls would be impervious to gas penetration, so that a costly refractory setting would be unnecessary.

In the design of the rear wall of the furnace, a "nose" would be incorporated at the furnace exit to insure good gas distribution. This wall would also form a three-row deep slag screen. The screen would be arrayed with a longitudinal spacing of 5-in. and a transverse spacing of 11-1/4-in. The boiler bank design would consist of 2-1/2-in. OD tubes, three rows deep on 7-1/2-in. spacing, and thirty-one elements across on 11-in. spacing, arrayed in an in-line configuration. The horizontal tubes would be 3-1/2-in. OD tubes, which would also be in-line on 5 x 5-in. centers. The loops would be supported from the front and rear

Philadelphia, Pa.: Feasibility Study

panel-walls of the second pass. Ample space would be provided in the design for the installation of sootblowers, if needed. The air heater would consist of 1,200 12 ft.-long, 2-1/2-in. OD tubes arranged in a 4-1/2-in. spaced in-line pattern. On the air-side, the gas flow would follow a three-pass, cross-flow path.

Feedwater from the refuse-fired economizers would be blended and sent over to the coal-fired steam generator at a temperature of 660°F., the drum pressure of which would be about 2,580 psig. Steam conditions would be 2,400 psig at 1000°F. reheat cycle. Excess air would be set at 18% and a steam flow of about 2.10×10^6 lb./hr. produced.

The steam generator fed by the two externally located economizers would be coal-fired rather than oil-fired, even though the latter is the prevalent mode of operation in the Philadelphia area. This is considered acceptable in that the system would incorporate gas cleaning equipment that would result in lower SO_2 emissions when firing high sulfur coal (approximately 3.0% S) than would be achievable by firing low sulfur oil (approximately 0.5% S) in existing, uncontrolled installations.

In most respects, the design of the coal-fired unit would be conventional. The main differences would be in the design layout of the heating surface. The heat absorbed in this unit would be largely accomplished by the superheater and reheater because of the use of the refuse-fired economizers. Some economizer surface would be included in the steam generator, however. This would be situated under the convection superheater. In conventional units a small section of the total economizer surface is usually located in this area, while the remainder is located to follow the parallel pass.

The furnace would have panel walls consisting of 3-in. OD tubes on 3-3/4-in. centers with continuous fins welded between the tubes. In the upper furnace, "wing" division walls would comprise a radiant superheater incorporating 2-in. OD tangent tubes. A parallel pass arrangement would be used in the second pass. Superheat temperature would be controlled by the firing rate and by spraying. Reheat temperature would be controlled by regulating the gas-flow with dampers.

The air pollution control equipment on the steam generator would consist of a wet scrubber system capable of handling the 745,000 acfm of flue gas calculated for this boiler. Exit flue gas temperature would be 300°F. The wet scrubber would remove both fly ash and SO_x at expected efficiencies of 99 and 90%, respectively. Sulfur oxide removal would be accomplished by liming the scrubber liquor in accordance with the Mitsubishi process. Gypsum recovery would not be attempted, however. The separated calcium sulfate would instead be discarded. A gas reheater would be included in the system to prevent the formation of stack plume.

The flue gas from the steam generator and the two economizers would not be intermixed at any point in the systems. Each would have its own air cleaning equipment; the economizers would, however, share a common stack. Each would deliver about 160,000 acfm of flue gas to its individual electrostatic precipitator, each of which would be rated at an efficiency of 99%. The estimated emission characteristics of the overall system can be seen in the table.

Estimated Stack Emission of 300 MW, Case 3 Power System

Stack	Fly Ash		SO_2	
	Pounds/Day	Grains/scf	Pounds/Day	Volume-Percent
Steam generator	4,392	0.042	26,352	0.021
Economizers	360	0.010	3,000	0.008
Total	4,752	–	29,352	–
Composite		0.034		0.018

The above estimates are based on several assumptions concerning fuel characteristics. It is assumed that the coal would have a sulfur and ash content of 3.0% S and 10.0%, respectively, and that 90% of each of these constituents would be entrained in the flue gas as SO_2 and fly ash, respectively. Refuse is assumed to contain 0.1% S, half of which would be converted to SO_2; it is also assumed that 24 lb. of fly ash would be formed for each ton of refuse fired.

A subject related to the above discussion is thermal pollution control. In the present analysis, cooling towers for restoring the original temperature of the riverine water discharged from the condenser system were not included. Such devices are not used by Philadelphia power stations, nor is it expected that they will become a requirement for future thermal plants. This situation apparently results from a number of factors, including good flow and mixing rates within each of the two major rivers there.

TABLE 4.10: ESTIMATED COSTS FOR A 300 MW COMBINED-FIRED TURBO-ELECTRIC PLANT

FPC Codes	Capital Costs	Cost, 10^6 $
310	Land and land rights	2.380
311	Structures and improvements	5.326
312	Boiler plant equipment	33.077
314	Turbine-generator equipment	13.431
315	Accessory electrical equipment	2.783
316	Misc. power plant equipment	0.398
	Air pollution control equipment	3.853
	Waste handling equipment	2.316
	Engineering and inspection	2.812
	Total Capital Cost	66.376

Annual Costs	
Annual Capital Cost	9.126
(Effective annualization rate = 13.75%)	
Water cost	0.006
Operating labor	0.723
Maintenance	1.411
Coal cost	5.301
Residue disposal	0.468
Total Annual Costs	17.035

Philadelphia, Pa.: Feasibility Study

Costs: Capital and annual costs for the 300 MW power plant are shown in Table 4.10. Those cost model modifications, which were discussed in connection with the steam plant and which are relevant here, have been applied. A fixed annual credit for power generated has not been used because of the current instability in fuel costs. In the Fuel Adjustment Clause of PE's tariff steam rates, fuel costs were increased 38% effective for the quarter starting February 1971. It is safe to estimate that fuel costs are in excess of $0.40/10^6$ Btu and will probably increase considerably in the very near future. On this basis, it can be said that refuse disposal costs for the proposed 300 MW plant would be something less than $1.75/ton and would probably shift to the asset side if fuel costs exceed $0.44/10^6$ Btu. The relationship of disposal costs to energy costs for the proposed plant are shown in Figure 4.9.

FIGURE 4.9: DISPOSAL COSTS FOR 300 MW PLANT AS A FUNCTION OF FUEL COST

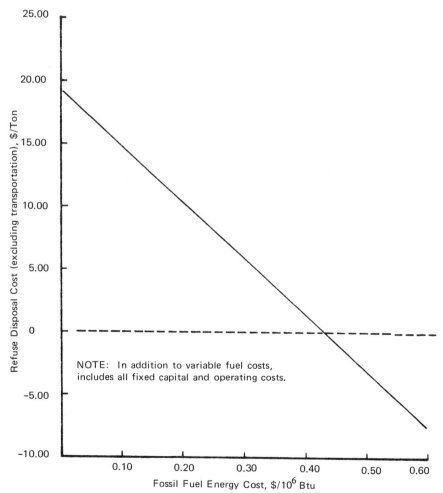

Site Selection and Refuse Transportation Costs: The logical location of the 300 MW plant would be on the Schuylkill or Delaware Rivers at some point northwesterly or northeasterly, respectively, of the northern boundaries of the Columbia and Fairhill Districts. As can be seen in Figure 4.1, two city incinerator sites can be considered. Because few other sites are known, these were studied to determine which would be optimum in terms of refuse transportation costs. The results of the vector analysis are summarized in the table.

Power Plant Refuse Transportation Data

District	Refuse Hauled, Ton-Miles/Year	
	N.W. Incinerator Site	N.E. Incinerator Site
Manayunk	971,300	1,380,700
Germantown	833,100	1,383,000
Logan	270,900	746,700
Frankford	1,685,800	317,000
Lower Tacony	492,700	147,500
Upper Tacony	4,049,100	2,350,800
	8,302,900	6,325,700

Based on the total amount of refuse handled and a haul cost of $0.25/ton-mile, the transportation cost would be $4.34 and $3.30/ton to the Northwest and Northeast Incinerators, respectively. Although the latter site is obviously to be preferred, the transportation cost is still unacceptably high, particularly in comparison with those of the existing incinerator system ($1.81/ton) and the proposed steam plant ($1.93/ton). The difference can largely be explained on the basis of the lower refuse production density that exists in the northern districts of Philadelphia, particularly Manayunk and the Taconys. An obvious solution would be to employ transfer stations in these particular districts, and possibly in Germantown as well. This should bring the costs down by about a fourth.

Cleveland, Ohio: Feasibility Study

This chapter is based on a report prepared by the Envirogenics Company for the EPA in 1971, the purpose of which was to develop design recommendations and procedures for the disposal of refuse, a low sulfur fuel, with heat recovery in utility grade boilers.

WASTE MANAGEMENT OPERATIONS

Two documents obtained from the City of Cleveland provided important information on refuse collection rates and population distributions. The first furnishes population statistics, by ward, for 1967, when it was estimated that the city population was 814,156. The 1960 and 1970 census figures are 876,050 and 750,903, respectively, a decrease rate of 1.55%/yr.

The ward structure of Cleveland is shown in Figure 5.1. The nine major ward-groupings demarcated represent a new waste management zoning being considered there. Proposed sites for future transfer stations are also shown on the map. The populations and population densities of these ward groups are shown in Table 5.1. It can be seen that regions of highest population densities are those surrounding downtown Cleveland and those along the Eastern end of the city.

Figure 5.2 shows the five (arbitrarily numbered) collection districts operating within Cleveland's Division of Waste Collection and Disposal. It will be noted that all five district yards (and offices) have been proposed as transfer stations. It will also be noted that two of the yards are not located in the districts they serve; in fact, the Harvard station is not even within the city limits. This of course results from land availability problems. The refuse collection rates for the five districts, based on 1969 data, are listed in Table 5.2.

Although the collection districts are not organized on the basis of ward boundaries, a geographic correlation can still be seen in that the heaviest concentrations of refuse production and ward population densities lie on the east end of town. This is fortuitous, in that the power plants located within

Cleveland are in or reasonably close to these areas of high population and refuse production. The two lakeside power generator sites shown in Figures 5.1 and 5.2 are, from west to east, owned by the City of Cleveland and by the Cleveland Electric Illuminating Co.

The collection rates shown in Table 5.2 below are regarded as being low. More accurate information was recieved for the year 1970, but this included totals only; these were broken down by districts.

The 1970 budget of Cleveland's Division of Waste Collection and Disposal (DWC&D) is approximately $12,500,000, second only to the budget of the Safety forces. Served on a weekly basis are 240,000 residential units, the cost of which is approximately $1.00/unit-wk. An estimated 395,000 tons of refuse was collected during 1970. The estimated per ton cost of disposing of Cleveland's refuse in 1970 was about $32.50, which included all services from pulling the containers to the curb to final disposal. Because this cost did not include such factors as capital improvements, amortization, interest, etc., the DWC&D regards $35.00/ton to be a realistic value.

TABLE 5.1: POPULATION DENSITY IN VARIOUS SECTIONS OF CLEVELAND (1967)

Ward Group*	Area, sq. mi.	Population	Pop. Density, persons/sq. mi.
A	9.4	136,132	14,480
B	6.8	59,075	8,688
C	6.5	127,407	19,600
D	5.5	71,398	12,981
E	5.6	73,043	13,043
F	9.7	87,948	9,067
G	7.7	75,462	9,800
H	14.5	90,336	6,230
I	7.6	93,305	12,277

*Letter designations are arbitrary.

TABLE 5.2: REFUSE COLLECTION RATES FOR CLEVELAND'S FIVE COLLECTION DISTRICTS

District	Area, Sq. Mi.	Refuse Collected TPD*	Refuse Collected TPD/Sq. Mi.	Estimated No. of Homes Served
1 - West Side	25.7	380	14.8	80,000
2 - 24th & Rockwell	13.1	140	10.7	60,000
3 - West 3rd Street	8.8	200	22.7	15,000
4 - Harvard	7.7	248	32.2	20,000
5 - Glenville	10.0	224	22.4	65,000

*Based on five-day work week.

Cleveland, Ohio: Feasibility Study

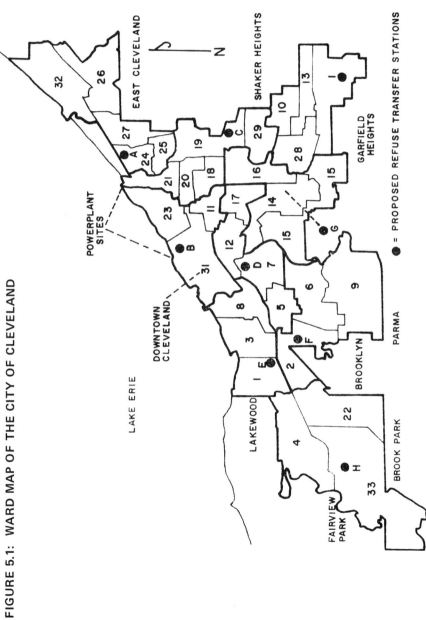

FIGURE 5.1: WARD MAP OF THE CITY OF CLEVELAND

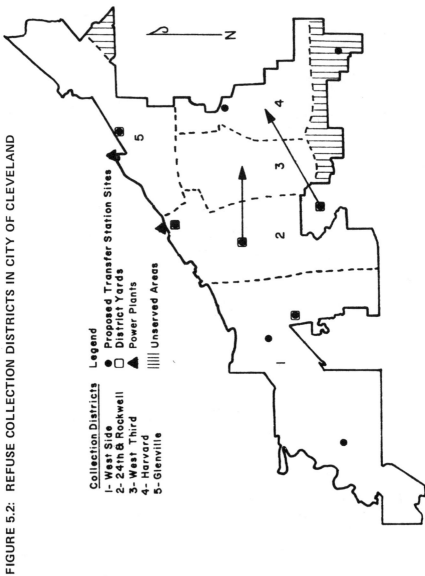

FIGURE 5.2: REFUSE COLLECTION DISTRICTS IN CITY OF CLEVELAND

The payroll of the division in 1970 exceeded 1,400 persons, although a reduction in forces through attrition is in progress to correct for certain archaic manning practices. The rolling stock of the division comprised 189 packer trucks, 28 flat-bed trucks, 48 passenger cars, and 21 specialized vehicles, including bulldozers, animal-carcass trucks, front-end loaders, etc. In general, the vehicle inventory is obsolescent, but new equipment is being actively sought.

Refuse handling within the five districts had been more or less standardized in recent years except as to method of disposal. Crews were first dispatched to move refuse containers to the curbs. Packer trucks next toured their routes and when filled proceeded individually, without intermediate transfer, either to the landfill site operated by the Rockside Hideaway Landfill, Inc., of Garfield Heights, Ohio, or to the Ridge Avenue incinerator.

The latter fires material collected only in the West Side and 24th and Rockwell districts, although both of these districts send more than 60% of their collections to the Rockside landfill. Formerly, there were two incinerators in operation in Cleveland. The older of these was phased out a few years ago as being too costly to rehabilitate. It is also planned that the 10 year old Ridge Avenue incinerator be eventually shut down and operations shifted to 100% landfill disposal.

Additional services provided by the DWC&D included the removal of furniture and large appliances from the regular routes, pickup of putrescibles from commercial and semicommercial establishments, and clean-up of dock areas and city streets. Special vehicles are used for these tasks.

At this time the DWC&D is moving forward in the modernization of its operations. This is vitally needed in view of the high cost of disposal and the fact that the Rockside landfill will soon be exhausted. In anticipation of this problem, bids are being sought for new landfill contracts wherein the operator would remove the refuse in trailers from close-in stations. This will be done by one of two options, as determined by cost analyses of the submitted bids. The first plan would involve the construction by the contractor of a receiving and storage plant.

It would not, however, involve any refuse grinding and the output from the storage bin would be compacted into trailers for outhaul to landfill rather than being conveyed to a furnace. The second plan would be based on the construction, by the city, of transfer stations at a minimum of two of the sites shown in Figures 5.1 and 5.2. There, the packer trucks would tip directly into the contractor's trailers.

When the selected plan goes into effect, the contractor will be expected to outhaul a guaranteed 300,000 tpy. The Ridge Avenue incinerator will continue to operate, although it is uncertain that the old refuse rate of 80,000 to 90,000 tpy will be maintained.

Another economic pressure recently felt by the DWC&D was a significant cut in the budget. This has necessitated an extensive reorganization of operations, which are still under evaluation. An outgrowth of this economy move has been a reduction in packer truck crews to three men and the abandonment of backyard trash pick-up.

Refuse Inventory and Composition Projections

Refuse Quantities: Data received from the City of Cleveland on the quantities of refuse disposed of during 1970 are shown in Table 5.3.

TABLE 5.3: QUANTITY OF WASTE DISPOSED OF IN CLEVELAND DURING 1970

Refuse Type	Cu. Yd./Yr.	Tons/Yr.	Tons/Day
Packer Truck			
to landfill	–	301,614	826
to incinerator	–	73,690	202
Bulky	219,330	17,766*	49*
Total		393,070	1,077

*Derived from a published specific volume value, where 1 cu. yd. of bulky refuse = 162 lbs.

Included in this refuse estimate was a predominating quantity of domestic material, together with street litter, dockside trash, and waste from institutional sources, hotels, and markets. The bulky waste constituted 4.5% of the total which is in good agreement with the 5% figure estimated earlier. Solid wastes from commercial and industrial sources, other than those mentioned, are handled by private companies.

The values shown above are regarded as being the most accurate yet obtained. The collection data examined for previous years, in fact, do appear to be on the low side, considering the high rates of annual increase they suggest. The earliest estimate on Cleveland's refuse collections was reported for 1966. The quantity estimated was 215,000 tpy, which would imply an increase of 14.8%/yr. to arrive at the 1970 figures. The quantity estimated for 1969 was 309,922 tpy, which is 21.7% less than the 1970 output. A future growth of 5%/yr. due to expansion in the southeast and southwest portions of the city was also predicted.

The 1.55%/yr. decrease in population should have slightly more than offset the expected 1.5%/yr. per capita increase in refuse production. Thus a more or less even collection rate should have been seen in Cleveland over the past decade. Because of the inaccuracies associated with older collection data and the prediction that Cleveland's population is soon expected to stabilize (remain constant), it was decided to assume that the 1970 data are accurate and project a growth rate of 1.5%/yr. over the next decade. On this basis, refuse rates were predicted to increase to 1,250 tpy by 1980. Allowing an additional 10% for commercial/industrial solid waste, the figure was set at 1,375 tpd.

Refuse Composition: The Martin-Marietta Co. has been conducting a study for EPA's Solid Waste Office. This has involved the compositional analyses of refuse collected at Orlando, Florida; Wichita Falls, Texas; and Cleveland, Ohio. The Project Manager, Mr. William Warren, kindly provided Envirogenics Co. the data which had been acquired in Cleveland. These analyses were made on refuse

Cleveland, Ohio: Feasibility Study

collected in selected areas serviced by trucks from the Ridge Avenue Station, the 24th and Rockwell Station, and the suburb of Olmstead.

The specific routes involved in the three areas were characterized, respectively, as consisting of: (1) public housing, (2) large, older single-family residences, many occupied by several families; and (3) single-family, middle-class residences. The averages obtained for these three samplings are shown in Table 5.4; data reported in the nationwide study are also given for comparison.

TABLE 5.4: AVERAGED COMPOSITIONAL DATA FOR CLEVELAND REFUSE

	Weight-Percent	
Constituent	Martin-Marietta, Cleveland	Nationwide
Garbage	18.3	20
Paper	32.3	38
Garden	11.9	12
Plastics	4.0	2
Metal	11.2	10
Glass	15.7	12
Residual	6.6	6
	100.0	100.0

The composition shown in Table 5.4 was then adjusted to reflect the presence of 10% commercial/industrial waste. Projections were then made for future compositions. The results are shown in Figure 5.3. From Figure 5.3 the composition of refuse in 1980 can be tabulated (Table 5.5). An ultimate analysis was then calculated for the composition shown in Table 5.5, the results of which are shown in Table 5.6.

TABLE 5.5: PROJECTED 1980 COMPOSITION OF CLEVELAND REFUSE

Constituent	Weight-Percent
Garbage	11.0
Paper	34.6
Garden	10.7
Plastics	6.7
Metal	8.5
Glass	12.6
Residual	5.9
Commercial/Industrial	10.0
	100.0

FIGURE 5.3: PROJECTED COMPOSITIONAL CHANGES IN CLEVELAND REFUSE

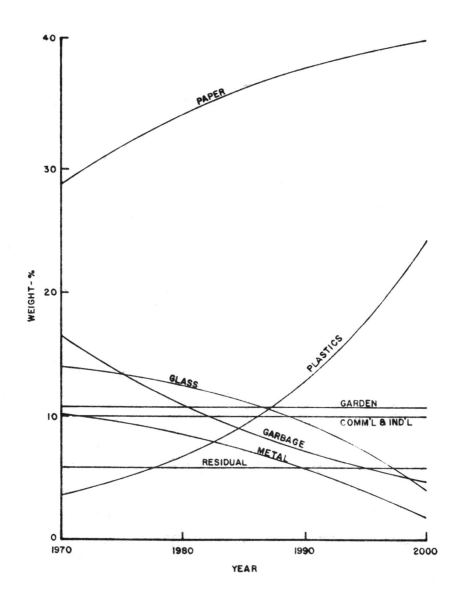

TABLE 5.6: PROJECTED 1980 ULTIMATE ANALYSIS OF CLEVELAND REFUSE

Constituent	Weight-Percent
H_2O	19.8
C	26.3
H	3.5
O	21.0
N	0.5
S	0.1
Inert	28.8
	100.0

Fuel Characteristics of Cleveland Refuse

Heating Value: Projections, based on the compositions discussed above and the constituent calorific values for refuse described in the earlier study, were also made for the heating values of Cleveland refuse. These are shown in Figure 5.4, where it can be seen that, by 1980, Cleveland's refuse will have a higher heating value (HHV) of about 4750 Btu's/lb.

Combustion Calculations: Combustion gas requirements, flue gas production rates, and steam generator efficiencies at exit flue gas temperatures of 450°, 500°, and 575°F. were then determined. These data are shown in Tables 5.7, 5.8 and 5.9. The three different temperatures correspond, respectively, to the flue gas exit temperatures of (1) the various earlier study refuse boiler designs, excluding Case 3, (2) the straight-refuse-fired district heat plant, and (3) the Case 3 refuse fired economizer. The efficiency data shown in Table 5.9 are graphed in Figure 5.5 to permit the extraction of values at other flue gas exit temperatures.

TABLE 5.7: COMBUSTION GAS REQUIREMENTS FOR CLEVELAND REFUSE COMPOSITION PROJECTED FOR 1980

	Combustion Gas Requirement, lb./lb. Refuse	
Constituent	Oxygen	Dry Air
C	0.701	3.021
H	0.280	1.207
S	0.001	0.004
Metal	0.007	0.030
	0.989	4.262
Oxygen	0.251	1.082
Total required, stoichiometric	0.738	3.180
Total required, 50% excess gas	1.107	4.770
Excess gas	0.369	1.590

FIGURE 5.4: PROJECTED CHANGE IN HEATING VALUE OF CLEVELAND REFUSE

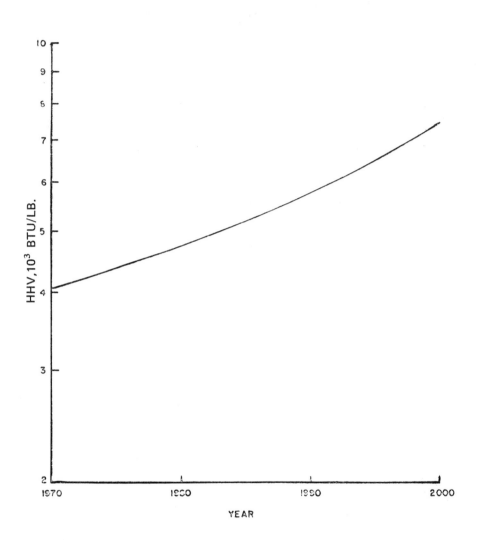

TABLE 5.8: PRODUCTS OF COMBUSTION OF THE REFUSE PROJECTED FOR CLEVELAND IN 1980

Constituent	Gas Formed per lb. Refuse		Volume-Percent Dry Basis
	Lb. Mol	Lb.	
CO_2	0.022	0.964	13.3
H_2O	(0.032)	(0.567)	–
Refuse H_2	0.018	0.315	–
Refuse H_2O	0.011	0.198	–
Combustion air	0.003	0.054	–
SO_2	<0.001	0.002	0.02
O_2 (excess)	0.012	0.369	7.3
N_2			
Total air N_2	0.131	3.666	79.4
Refuse N_2	<0.001	0.004	
Total flue gas (wet)	0.199	5.572	
Total flue gas (dry)	0.167	5.005	

TABLE 5.9: EFFICIENCY OF STEAM GENERATOR FIRING CLEVELAND REFUSE OF COMPOSITION PROJECTED FOR 1980

Item	Fuel Value Heat Losses at Various Flue Gas Exit Temperatures, Btu/lb. of Fuel (% HHV)		
	450°F.	500°F.	575°F.
Dry gas	444 (9.35)	505 (10.63)	595 (12.53)
H_2O in refuse	241 (5.07)	245 (5.16)	252 (5.31)
From H_2 combustion	383 (8.06)	390 (8.21)	401 (8.44)
H_2O in air	11 (0.23)	13 (0.27)	15 (0.32)
Unburned gas	4 (0.08)	4 (0.08)	4 (0.08)
Unburned residue	110 (2.32)	110 (2.32)	110 (2.32)
Sensible heat, residue	46 (0.97)	46 (0.97)	46 (0.97)
Unburned fly ash	41 (0.86)	41 (0.86)	41 (0.86)
Sensible heat in fly ash	5 (0.09)	5 (0.11)	6 (0.13)
Subtotal	1,285 (27.03)	1,359 (28.61)	1,470 (30.96)
Radiation	(0.20)	(0.20)	(0.20)
Unmeasured	(0.50)	(0.50)	(0.50)
Manufacturer's margin	(1.00)	(1.00)	(1.00)
Total % heat loss	28.73	30.31	32.66
Steam gen. efficiency	71.27	69.69	67.34

NOTE: Fuel value (HHV) = 4,750 Btu/lb.;
Combustion air inlet temperature = 80°F. (60% RH)

FIGURE 5.5: EFFICIENCIES AT VARIOUS FLUE GAS EXIT TEMPERATURES OF A STEAM GENERATOR FIRING PROJECTED 1980 CLEVELAND REFUSE

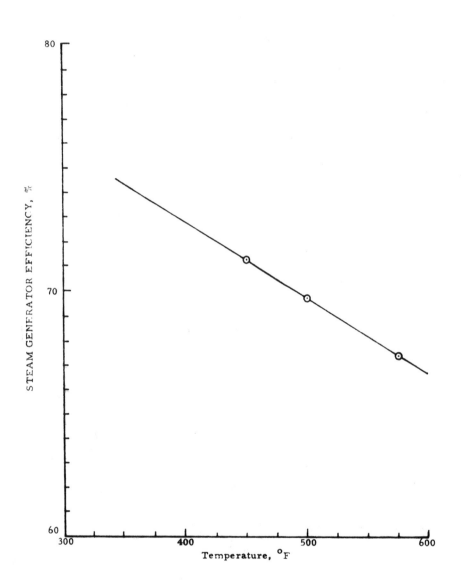

Cleveland, Ohio: Feasibility Study

Utility Steam Generation Operations

Municipally-Owned Boilers: The facilities owned by the City of Cleveland consist of six units, five of which are wet bottom or slagging furnaces. The sixth and newest unit was installed within the past few years. It has a pressurized furnace and is of 75 MW capacity.

The other five units are Foster Wheeler boilers, three of which were placed in service in the early 40's. The other two were commissioned in 1956. All five fire pulverized coal of low ash fusion temperature from horizontally aligned burners. The three old boilers are identical, each being designed to generate 300,000 lb./hr. of steam at a heat release rate of 26,100 Btu/ft.^{3}. They are of course refractory-walled furnaces, the monowall construction being of more recent invention. Retrofit of slag-tap boilers to refuse or combined (refuse plus fossil fuel) firing is impractical because of the costly modifications that would be necessary. Pressurized-furnace boilers are also poor candidates for retrofit because of the difficulty of providing a workable gas-seal on the charging chute.

Cleveland Electric Illuminating Co. (CEI) - Owned Boilers: Three plants are operated by the CEI which are within reasonable distance of the municipal refuse collection system. These are the Lakeshore power station on East 70th Street and the Cleveland Memorial Shoreway, the Canal Road steam plant beneath Eagle Avenue bridge, and the East 20th Street steam plant between Hamilton and Lakeside Avenues. Five units are operated at the Lakeshore Station, four of which are of the continuous slag-tap type and thus incompatible with refuse firing. The fifth unit is a 250 MW dry-bottom boiler which is fueled with tangentially-fired, pulverized coal. It is of recent enough vintage (1959) that it would not be considered for retrofit treatment.

The steam plant at East 20th Street comprises six units, all of which were commissioned before 1930. The location of the plant is in a highly congested area such that it would not be suitable for refuse processing.

The Canal Road plant contains five boilers, all of which were installed between 1948 and 1950. Each is equipped with forced draft chain grate stokers and is capable of generating up to 150,000 lb./hr. of steam. Because of their small sizes, it is doubtful tht retrofit would be practical.

Effect of Firing Refuse on Pollution Burden: The benefits that firing the refuse as a substitute for fossil fuel would have on the Cleveland air pollution situation is difficult to assess because of the rather transitory nature of present utility fuel-use practices.

It was decided, somewhat arbitrarily, that benefit estimates would best be derived based on the assumption that a coal containing about 1% sulfur and 10% ash would be the fuel that would be partially replaced by the use of refuse. Because gas cleaning equipment is not yet generally used throughout Cleveland's coal-burning utility stations, particulate emission changes were calculated on the basis of both zero and 99% control. As in most of the nation, no APC equipment for sulfur oxide emission abatement is in use in the Cleveland area, although plans for one such system are being considered. For the present analysis, however, zero SO_2 control was assumed.

From these bases, the comparative emissions of refuse and an equivalent (energy-wise) amount of coal were calculated. It was found that coal would produce about twice as much particulate loading and 4 1/2 times as much SO_2 in stack gas then would refuse.

The values assigned to the system variables are given in Table 5.10. They are based on data presented in the preceding text and typical air pollution data provided from the earlier study.

TABLE 5.10: FACTORS ASSUMED FOR AIR POLLUTION CALCULATIONS

Fuel	HHV, Btu/lb.	Boiler Efficiency, %	Flue Gas Loading, - lb./ton Fuel Fired -	
			Particulates	SO_2
Refuse	4,100	70	24	2.3
Coal	12,000	85	160	19.0

On the basis of the above factors, it was estimated that 3.55 lb. of refuse would be required to produce the same working fluid enthalpy as 1 lb. of coal. Thus for every ton of refuse fired in substitution for coal, the flue gas exiting the air heater would contain 21 lbs. less particulate matter and 3 lbs. less SO_2. If an electrostatic precipitator of 99% efficiency were used, the particulate reduction in the stack gas would only amount to 0.21 lb. per ton of refuse fired in substitution for coal. The precipitator would of course have no significant effect on SO_2 levels.

If all the refuse now collected in the City of Cleveland (393,000 tpy) were fired in substitution for coal (1% S and 10% ash), the SO_2 atmospheric burden would be reduced by about 590 tpy. Air-borne dust would be reduced by about 4,125 tpy if no APC systems were involved, but only by one hundredth that amount if gas cleaning equipment of 99% efficiency were in use by the hypothetical coal/refuse-fired plants. As found in the Philadelphia analysis, the air pollution abatement benefits available from firing refuse in lieu of fossil fuel are not great.

PRELIMINARY PLANNING RECOMMENDATIONS

Overview

Undertaking the construction of one or more refuse-firing steam generators in a large metropolitan area (LMA) such as Cleveland is constrained by a number of factors. Obviously, the systems selected must be located near the refuse collection network to minimize transportation costs. Similarly, the operation of a single installation, capable of handling all of the City's refuse, should be avoided. A single disposal plant would necessarily be nonoptimum in terms of city-wide accessibility and, too, such an operation would doubtless promote severe traffic problems in the immediate vicinity of the plant.

Another key factor which must be considered is that such plants must be situated near their service interfaces. A turboelectric facility should be within easy reach of the electrical grid system and of copious quantities of cooling water. A process steam or district heating plant must be as close as possible to the steam service lines because easements of any distance are often difficult to arrange. Finally, and perhaps most importantly, the planning interests of the

Cleveland, Ohio: Feasibility Study

utility companies, the city managers, and potential large users of the product (e.g., process steam) must be served.

In the study of the Cleveland resource system, three aspects of the local energy demand situation were considered. First, as in any LMA, electrical power demand is increasing at a vigorous rate. For a city of Cleveland's size, as much as 700 MW additional capacity per decade may be required. Secondly, heating steam demand, though much more static than the demand for electrical energy, will be difficult to satisfy with the aging inventory of boilers now in service. Thirdly, the East Ohio Gas Co. has an interest in the concept of firing refuse to generate process steam (2 to 3×10^5 lb./hr.) for the Republic Steel Plant in Cleveland.

Unfortunately, the last item became known only late in the program and a working arrangement between Envirogenics and the Cleveland principals that would be consistent with scheduled commitments could not be arranged.

Thus in the preliminary design work, the decision was made to consider a refuse reduction arrangement which would include a district heating boiler and a turboelectric unit. The combination of these two "strategically separated" facilities would offer the capability of consuming all of the refuse generated in Cleveland by the year 1980.

Refuse-Fired District Heating Plant

Site Selection: The district heating system operated by CEI is shown in Figure 5.6. The only sites recognized for possible construction of a refuse-fired facility were that of the existing steam plant at Canal Road, the grounds of the now-defunct 3rd Avenue municipal incinerator, and the yard of Sohio's asphalt and steam plant between Broadway and East 34th Street.

The Canal Road plant is obviously the best choice because it is already an element of the existing steam system. The plant lay-out does not offer adequate room for waste handling, but some land is available across the road from the plant. The 3rd Avenue municipal incinerator is only 0.7 miles from the Canal Road steam plant and the existing steam network.

However, the construction of a pipeline does not appear economically attractive. Any steam flowing between the two plants would have to cross the Cuyahoga River, the Inner Belt Freeway, and a number of railroad lines. The Sohio process plant is a little farther (1.2 miles) from the Canal Road plant and is also separated by freeway and railroad systems. It is, however, on the right side of the Cuyahoga and can be considered the second best choice. Purchase of the Sohio steam plant by the CEI is a possibility.

In examining the Canal Road plant layout (see Figure 5.7), a possible accommodation for a refuse-firing unit has been recognized. The existing 5-unit plant was designed with provision for the installation of a 6th unit, the position of which is the closest to the coal storage facility. The latter consists of a concrete walled, 30-ft. deep pit having a floor area of about 1/3 acre. Coal feeding, however, is normally done directly from hopper bottom railroad cars over the track hoppers.

When this process is interrupted, then charging is done from the pit, but usually only from that portion of the pit under the trestle into which coal cars can discharge. A front end loader moves the coal from this area to a loading

90   Energy from Solid Waste

hopper which communicates with a conveyor system. The latter carries the feed to coal-crushers and then to storage hoppers which feed the chain stokers.

A major portion of the coal-pit contents is held in reserve for emergencies and has so been held for many decades. If this coal could be stockpiled elsewhere, then the pit could be partitioned and a bulk of its volume used for refuse storage without interfering with normal coaling operations.

FIGURE 5.6: CLEVELAND'S DISTRICT HEATING SYSTEM

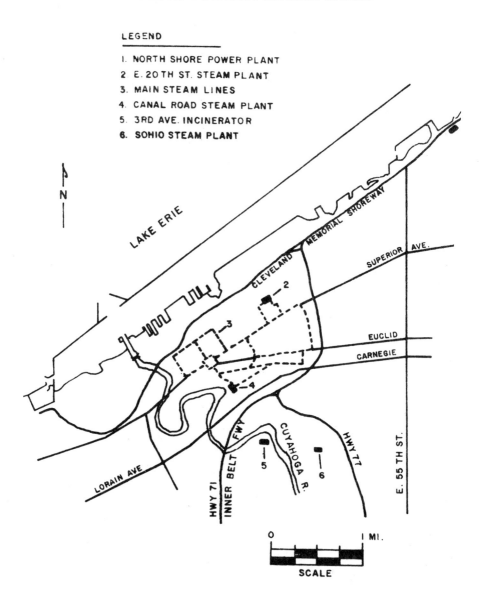

Cleveland, Ohio: Feasibility Study

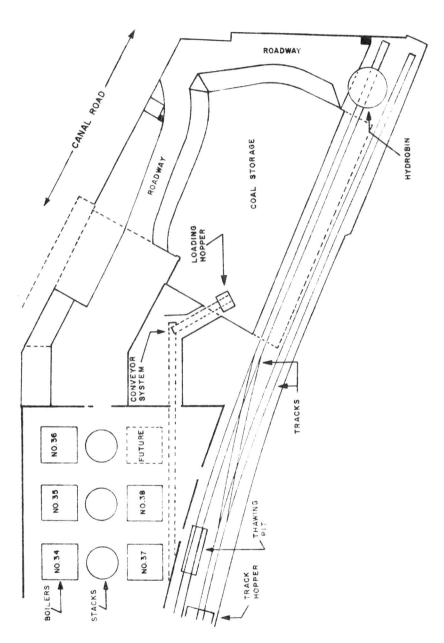

FIGURE 5.7: CANAL ROAD STEAM PLANT

The steam from the existing CEI steam plants is sent out at 170 psig, equivalent to an enthalpy of about 1,230 Btu/lb. The minimum summer load requires steam production ranging between 90,000 and 120,000 lb./hr. of steam. The former value was selected as the rating for the refuse firing unit. This would permit the unit to dispose of refuse collected in designated Cleveland districts at a more or less steady rate all year around. It would also permit the East 20th Street plant to feed some steam into the loop during minimum demand periods and thus minimize frictional line-losses. The duty of the new unit would be 0.105×10^9 Btu/hr. Using the fuel value projected for Cleveland refuse in 1980 (4,750 Btu/lb.) and an efficiency calculated to be 69.7%, a refuse rate of about 400 tpd would be required.

Over-the-Fence Arrangement: The over-the-fence installation would be essentially a self-sufficient system. It would consist of a single boiler, the enclosure of which would include a tipping-pit. The overall system design and operation would be basically the same as that described in the previous section except of course smaller in scale. The characteristics of the proposed Canal Road steam plant are summarized in Table 5.11.

As seen, the storage pit would be sized to provide six tipping stations. Based on an average packer truck load of four tons (15 cu. yds.) and a discharge cycle time of four minutes, an average dumping rate of 360 tph would be expected. The maximum receiving rate during peak traffic hours should not, however, exceed 150 tph. Thus, truck queuing is not probable. Weigh-in would be handled by a single, automated scale. The access streets serving the site are rather narrow and require improvement. Because of this, it was felt that oversized refuse should not be handled at this plant.

TABLE 5.11: SYSTEM CHARACTERISTICS OF CANAL ROAD STEAM PLANT (REFUSE-FIRED)

Steam Specifications

Production, lb./hr.	90,000
Sendout, lb./hr.	75,000
Pressure, psig	170
Temperature, °F.	425

Boiler Specifications

Refuse rate, tpd	400
Efficiency, %	69.7
Duty, 10^9 Btu/hr.	0.105
Flue gas exit temperature, °F.	500
Exit flue gas volume, ACFM	76,600
Feed water temperature, °F.	220
Plant factor, %	80

Refuse Handling Facilities Specifications

Tipping pit dimensions, ft.	100 (l) x 40 (w) x 50 (d)
Number of bridge cranes	2 (60 ft. long)
Grapple capacity, cu. yd.	3
Number of tipping stations	6

Cleveland, Ohio: Feasibility Study

Costs for the overall system have been derived utilizing the methodology outlined in the earlier study. In these calculations, the present system was considered to be equivalent to a 10 MW turboelectric system in solving certain of the cost functions. Because the base date of the various cost formulas is June 1969, capital costs have been increased by 10%/yr. for a period of two years, bringing the base date to June 1971. The results of this costing are shown in Table 5.12.

Disposal costs have been derived in two ways. In Method A, it is assumed that the new, refuse-firing steam generator would be added to the system inventory to provide needed capacity. That is, it would replace a unit that had to be retired or would provide a needed increase in the total production of the existing system.

TABLE 5.12: ESTIMATED COSTS FOR THE OVER-THE-FENCE, REFUSE-FIRED STEAM BOILER

Capital Costs

FPC Codes	Description	Cost, 10^6 $
310	Land and land rights (10 acres)	0.860
311	Structures and improvements	0.348
312	Boiler plant equipment	2.360
315	Accessory electrical equipment	0.202
316	Misc. power plant equipment	0.097
	Air pollution control equipment	0.189
	Waste handling equipment	0.914
	Engineering and inspection	<u>0.705</u>
	Total Capital Cost	5.675

Annual Costs

Annual capital cost, 10^6 $	0.828
(Effective annualization rate = 14.6%)	
Water cost, 10^6 $	0.038
Operating labor, 10^6 $	0.455
Maintenance, 10^6 $	0.096
Residue disposal, 10^6 $	<u>0.090</u>
Total Annual Costs, 10^6 $	1.507

	A	B
Annual credit, 10^6 $	0.552	0.519
Waste burned, 10^3 tpy	117	117
Disposal cost, $/ton	8.16	8.45

(A) Based on revenues for steam generated.
(B) Based on coal, labor and maintenance costs for operating existing steam generator of equivalent capacity.

In either of these situations, it would then be acceptable to apply the value of the steam generated (based on conventional plant operation) against the annualized capital and O&M costs of the new plant. If, on the other hand, the addition of this base load plant were made to an existing steam plant arrangement that was already capable of handling the demand, a service displacement would occur. Some or all of the conventional units would have to be operated at reduced plant factors in order to permit the refuse-fired installation to operate at full load. In this case, annual credits (Method B) should include only the costs of fossil fuel, labor, and maintenance for the conventional plants partly or totally displaced by the new plant.

The credit used for the Column A costing was derived using a steam value of $1.05/10^3$ lb., based on present operating costs. For Column B, a coal-fired steam generator efficiency of 83% and energy cost of $0.31/10^6$ were used in determining the coal credit. Labor and maintenance savings were assumed to be the same as the corresponding costs for operating the refuse-fired plant, except that those cost items dealing with refuse handling were deleted. This adjustment, incidentally, showed that the labor costs of the refuse-fired plant would be 67% higher than those of its fossil fuel equivalent.

A drawback to the plant layout just described is the condition of the land adjacent to the Canal Road Plant. It is poorly shaped and contoured, offers a marginal amount of area, and has a transformer building located on the most important section of the plot. A detailed civil engineering analysis of the property would be required to verify the feasibility of utilizing this real estate.

On-Site Arrangement: An alternate approach would be to utilize the vacant foundation in the Canal Road Plant for the installation of a refuse-fired boiler (see Figure 5.7). This would require that at least a portion of the coal pit would be available for refuse storage.

The floor area of the pit is 14,100 sq. ft. and the walls are 30 ft. high except around the trestle. This represents a storage capacity of over 15,000 cu. yds. which is well in excess of the maximum refuse storage volume requirement of 7,000 cu. yds. If the plant were converted to oil firing, the existing coal conveyor system, including the loading hopper on the floor of the pit, would be of no value for refuse charging. The existing components are too narrow and conversion to the 10-ft. felt width needed would be excessively costly. A new charging arrangement would have to be installed.

Direct tipping into the pit also will not be practical. The two ramped-roadways are too narrow and the pit wall extends from four to nine feet above them at the highest and lowest elevations of the ramps, respectively. The most economical solution to this problem is to install a four-station tipping pit on the property opposite the plant on Canal Road. An enclosed conveyor system would then be used to bring the material from the tipping pit, over Canal Road, and into the storage structure. The overall arrangement is shown conceptually in Figure 5.8.

An expensive pit modification will be the installation of a covering structure. Except for a flat portion under the trestle, it would be an arched configuration, supported by columns standing within the pit; the roof would extend just to the walls at the edge of the pit. In the interest of odor control, the structure would be fitted with duct work so that the combustion air for the refuse-fired furnace could be drawn from the pit.

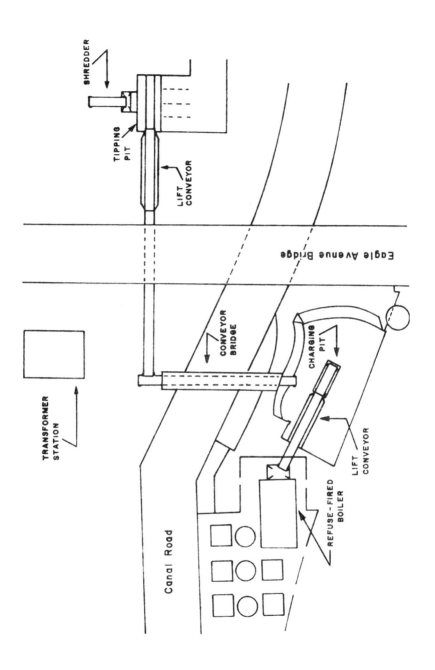

FIGURE 5.8: PROPOSED MODIFICATION OF CANAL ROAD PLANT

Because of its lower volumetric heat-release rate, the boiler itself would be much larger than the units now fired at Canal Road, such that the existing foundation site would not be adequate. As shown in Figure 5.8 the boiler would therefore be laid out so as to project out toward the refuse pit. This would require the removal of a portion of the boiler house wall, and the erection of a new ell on the building to house the new boiler. The charging hopper of the boiler would be sealed off from the rest of the building so that the refuse odors would not permeate into the plant. The boiler design features would be essentially the same as those described for the over-the-fence plant, except that stoking would be by conveyor and ram injector.

Preliminary costs for this system were derived and are presented in Table 5.13. It can be seen that both capital and annual costs are lower than for the over-the-fence plant. This is because the operating labor costs will be somewhat lower by having the plant on site and because less land will have to be purchased. In this costing, bulldozer equipment sufficient to support two operators per shift have been included even though such equipment is now being operated.

TABLE 5.13: ESTIMATED COSTS FOR REFUSE-FIRED STEAM BOILER INSTALLED ON-SITE AT CANAL ROAD PLANT

Capital Costs

FPC Codes	Description	Cost, 10^6 $
310	Land and land rights (5.5 acres)	0.471
311	Structures and improvements	0.400
312	Boiler plant equipment	2.311
315	Accessory electrical equipment	0.202
316	Misc. power plant equipment	0.050
	Air pollution control equipment	0.189
	Waste handling equipment	1.022
	Engineering and inspection	0.660
	Total capital cost	5.305

Annual Costs

Annual capital cost, 10^6 $		0.778
(Effective annualization rate = 14.6%)		
Water cost, 10^6 $		0.038
Operating labor, 10^6 $		0.416
Maintenance, 10^6 $		0.133
Residue disposal, 10^6 $		0.090
Total Annual Costs, 10^6 $		1.455

	A	B
Annual credit, 10^6 $	0.552	0.519
Waste burned, 10^3 tpy	117	117
Disposal cost, $/ton	7.72	8.00

(A) Based on revenues for steam generated.
(B) Based on coal, labor and maintenance costs for operating existing steam generator of equivalent capacity.

Cleveland, Ohio: Feasibility Study

Air Pollution Control (APC) Equipment: In either of the possible plant layouts, the APC system would be the same. A single electrostatic precipitator having a dust removal efficiency of 99% would be used. The system would be sized to handle a gas throughput of 76,600 acfm based on an inlet temperature of 500°F. Because only refuse (sulfur content approximately 0.1% S) would normally be fired in the furnace, provision for SO_2 removal is not considered necessary.

Using the same emission factors assumed for the Philadelphia study (Chapter 4), it can be estimated that the present unit would emit about 96 lb./day fly ash. This corresponds to a stack gas loading of 0.011 g./scf. The SO_2 output would be about 800 lb./day; this corresponds to a stack gas concentration of 0.008 volume percent.

Refuse-Fired Turboelectric Plant

Size Specification: Whichever version of the steam plant discussed in the previous sections were built, it would be capable of handling only slightly more than 1/4 of the total refuse projected to be collected in Cleveland in 1980. Disposal of an additional 975 tpd will probably be necessary. This quantity is about right for a 200 MW Case (936 tpd) system, deriving 16.6% of its heat input from refuse. Adding this much power capacity to the existing CEI inventory does not appear to pose any problem, considering the growing power demand in Cleveland. The Case 3 design, as discussed in the earlier study is considered to be optimum for this power rating among the many conventional and advanced (e.g., suspension firing) designs analyzed for cost effectiveness.

A question concerning plant sizing was whether to provide additional capacity to accommodate refuse production growth beyond 1980. This was not done in the present case, because the possibility also exists in Cleveland for the erection of a third refuse-fired plant sized for the production of 2 to 3×10^5 lb./hr. of process (steel mill) steam.

Site Selection: Studies of possible sites for the plant revealed that the plant would probably have to be located on the shores of Lake Erie. The Cuyahoga River is too contaminated to be considered even as a source of cooling water. Along the lake front, however, relatively few sites, which are within reasonable reach of the waste collection system, appear to be available. The best approach would therefore be to install the new boiler at the existing Lake Shore Plant. This already accommodates five units, but has provisions, including an empty turbine room, for expanded capacity. Installation of a Case 3 system could be accomplished by locating the fossil fuel steam generator on the "future-site" provided and locating the refuse-fired economizer on adjacent property. The general plant layout is shown in Figure 5.9.

System Characteristics: A 200 MW version would be comprised of one coal-fired steam generator and one refuse-fired economizer. The tipping pit would be 150 ft. long, 40 ft. wide, and 60 ft. deep. Two cranes mounted on 60 ft. bridges and equipped with 5 cu. yd. grapples would be provided. One crane would be used for standby service except during the day shift on week days, when it would be manned and used to mix and arrange the pit contents.

The general design and operational characteristics of the system would be very similar to the Case 3 system equipped with two refuse-fired economizers described in Chapter 4. Specific descriptive information on the present system is itemized in Table 5.14.

FIGURE 5.9: LAKE SHORE PLANT LAYOUT

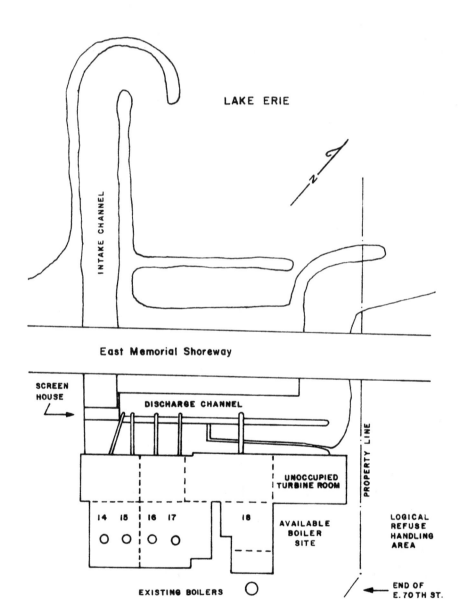

TABLE 5.14: CHARACTERISTICS OF 200 MW, COMBINATION-FIRED SYSTEM FOR THE LAKE SHORE PLANT

Item	Economizer	Steam Generator
Fuel rate, tpd	936 (Refuse)	1,752 (Coal)
Excess air, %	50	18
Flue gas exit temperaure, °F.	575	300
Flue gas volume, ACFM	195,000	570,000
Unit efficiency, %	67.3	87.0
Design fuel value, Btu/lb.	4,750	12,022
Duty, 10^9 Btu/hr.	0.249	1.527
Feedwater temperature, °F.	440	620 (1960 psig)
Steam conditions, psig/°F./°F.	–	1,800/1000/1000
Steam flow, 10^6 lb./hr.	–	1.430
Number of turbines	–	1
Plant factor, %	80	80

Because of its more accessible location, the Lake Shore system would be assigned the task of shredding and firing oversized refuse. A single 700 hp. hammermill could reduce such items at a rate of at least 30 tph. The city's entire output of such refuse is only about 50 tpd at the present time and would probably increase to only 75 tpd by 1980. Thus the mill would only be used a few days each week.

A special design requirement for this proposed plant would be low stack height. The present Lake Shore site is situated in an aircraft lane and CEI has already been requested to shorten the boiler stacks already erected on the property. There is, in fact, a possibility thst the site may be condemned for the above reason.

The addition of a combination-fired system would not, however, complicate the stack hazard problem. Being equipped with high efficiency gas cleaning systems, there would be no need to rely on high stacks to disperse pollutants. The wet scrubber system on the coal-fired steam generator would be sized to handle the 570,000 acfm of flue gas calculated for this boiler based on an exit flue gas temperature of 300°F. The wet scrubber would remove both fly ash and SO_x. The latter removal would be accomplished by liming the scrubber liquor in accordance with the Mitsubishi process. Gypsum recovery would not be attempted, however. The separated calcium sulfate would instead be discarded. Because of the inclusion of this system, use of a low sulfur coal would be unnecessary.

The refuse-fired economizer would be equipped with an electrostatic precipitator having a dust-removal efficiency of 99%. It would be sized to handle a gas throughput of 195,000 acfm based on an inlet gas temperature of 575°F. Following the gas cleaning stages, the exit gases from the two furnaces would be blended and discharged through a common stack. Because of the comparatively high temperature of the economizer flue gas, it would be unnecessary to reheat the flue gas exiting the wet scrubber of the steam generator. Using the same emission factors observed for the Philadelphia analysis (Chapter 4), the following emission estimates can be tabulated (Table 5.15).

TABLE 5.15: ESTIMATED STACK EMISSIONS OF 200 MW CASE 3 POWER SYSTEM

Source	Fly Ash		SO_2	
	Lbs./Day	Grains/SCF	Lbs./Day	Volume-Percent
Steam generator	3,154	0.039	18,922	0.020
Economizer	224	0.012	1,872	0.008
Combined stack	3,378	0.033	20,794	0.017

Costs for this system have been derived as shown in Table 5.16 and in accordance with the procedures described earlier in this report. In computing the annual credit for power generated, a parallel costing was performed for a conventionally-fired 200 MW unit to determine annual costs and thus, production costs for electricity, based on today's capital and operating expenses. Unlike Philadelphia, considerable amounts of coal are being fired in Cleveland such that fuel costs are more stable. Because of this, a fixed fuel cost of $0.31/$10^6$ Btu could be used in the calculations. It can be seen from the cost data that the disposal cost for refuse, exclusive of transportation, is considerably more attractive than those derived for the district heating plant.

TABLE 5.16: ESTIMATED COSTS FOR REFUSE-FIRED, TURBO-ELECTRIC SYSTEM INSTALLED AT THE LAKESHORE STATION

Capital Costs

FPC Codes	Description	Cost, 10^6 $
310	Land and land rights	1.040
311	Structures and improvements	3.343
312	Boiler plant equipment	23.663
314	Turbine-generator equipment	10.582
315	Accessory electrical equipment	1.682
316	Misc. power plant equipment	0.384
	Air pollution control equipment	2.550
	Waste handling equipment	1.362
	Engineering and inspection	2.268
	Total Capital Cost	46.874

Annual Costs

Annual capital cost, 10^6 $	6.844
(Effective annualization rate = 14.6%)	
Operating labor, 10^6 $	0.551
Maintenance, 10^6 $	1.107
Coal cost, 10^6 $	3.813
Residue disposal, 10^6 $	0.249
Total Annual Costs, 10^6 $	12.564
Annual credit for power generated, 10^6 $	11.849
Quantity of waste burned, 10^3 ton/yr.	273
Disposal cost, $/ton	2.62

Transportation Costs

An important cost element in disposing of refuse is that associated with the movement of packer trucks, once they are loaded, from their collection routes to the disposal site. If this cost increases in shifting from present disposal methods to the steam generator approach, appropriate operating adjustments will be required. Either the rolling stock and the number of collection routes must be increased or transfer stations will have to be incorporated within the system. In performing the cost analysis, a number of assumptions were necessarily made. These are discussed in the following sections.

Tonnage Hauled: It was assumed that the amount of collected refuse generated daily would be that projected for 1980 (1,375 tpd). It was further assumed that the relative distribution of this production among the five collection districts would be the same as that recorded for the year 1969. The values thus derived are shown in Table 5.17. It was also assumed that the refuse production densities within each district would be uniform.

TABLE 5.17: CLEVELAND REFUSE PRODUCTION PROJECTED FOR 1980 BY DISTRICTS

Collection District	Collected Refuse Generation Rate, tpd	Percentage of Total
(1) West Side	439	31.9
(2) 24th and Rockwell	161	11.7
(3) West 3rd	231	16.8
(4) Harvard	286	20.8
(5) Glenville	258	18.8
Total	1,375	100.0

Disposal Sites: In attempting to compare transportation costs of the present methods of disposal with those associated with steam plant operation, the disposal sites must be identified. This is difficult to do in that the landfill area now in use will doubtless be exhausted before the end of the decade and other sites, probably more distant from the collection system, will have been put into operation. Because the location of future landfill sites is unknown at the present time, it was necessary to assume that the Rockside landfill would still be in use in 1980. Thus the transportation costs derived for landfill disposal are probably low.

The other disposal sites were assumed to be the Ridge Road Incinerator, and refuse-firing boilers at the Canal Road steam plant and the Lakeshore power station. The areas assumed to be served by these facilities are itemized in Table 5.18.

Travel Distance Derivations: Weight-distance vectors were derived as explained in Chapter 4.

Results: Using a base cost of $0.20/ton-mile, it was found that haulage to the existing disposal sites would average out at $2.58/ton. The information used to derive this value is shown in Table 5.19. The cost would be somewhat lower if current refuse production quantities were used, since then the fraction of the

TABLE 5.18: REFUSE INPUT AREAS ASSUMED FOR TRANSPORTATION COST ANALYSIS

Disposal Site	Area Served	Refuse Input, tpd
Present Disposal Methods:		
Ridge Road Incinerator	District 1 east of West 117th Street	350
Rockside Landfill	District 1 west of West 117th Street, plus all other districts	1,025
Disposal in Steam Generators:		
Canal Road Plant	District 1 and District 2 south of Denison Avenue	400
Lakeshore Plant	District 2 north of Denison Avenue plus Districts 3, 4 and 5	975*

*This is slightly higher than the actual capacity (936 tpd) of the plant. This was done to permit direct transportation cost comparisons of the two disposal methods. This adjustment does not bias the results.

TABLE 5.19: REFUSE TRANSPORTATION COST FACTORS FOR EXISTING CLEVELAND DISPOSAL SITES

Site	Avg. Direct Haul Distance, mi.	Refuse Transported, tpd	Average Round Trip Mileage*	Ton-Miles
Ridge Avenue Incinerator:				
District 1, east of West 117th Street	1.3	350	3.3	1,138
Rockside Landfill:				
District 1, west of West 117th Street	9.0	89	22.5	2,003
District 2	5.8	161	14.5	2,335
District 3	5.1	231	12.8	2,957
District 4	4.8	286	12.0	3,432
District 5	9.1	258	22.8	5,882
Total		1,375		17,747

$$\text{Transportation cost} = \frac{\$0.20 \times 17{,}747}{1{,}375} = \$2.58/\text{ton}$$

*Includes 25% increase for direct travel patterns.

Cleveland, Ohio: Feasibility Study

total refuse handled at the more centralized Ridge Road incinerator would arithmetically increase. Regardless of which base year is observed, however, the cost would be high enough to warrant serious consideration of the use of transfer stations, as is now being done.

Costs were similarly derived for transporting refuse to the proposed steam generators. The data are summarized in Table 5.20. The cost of $2.17/ton is significantly lower than that associated with haulage to the existing system ($2.58/ton) and would become even more favorable as new, more distant landfill sites are brought into operation.

By way of breakdown, transportation cost to the proposed Canal Road plant would be $2.80/ton, while that to the Lakeshore Station would be only $1.91/ton. This would suggest that transfer stations also be considered in connection with the operation of the refuse-fired boiler at Canal Road. Referring to Figure 5.1, the most westerly plus one of the two other transfer station sites proposed for District 1 would appear to be suitable.

TABLE 5.20: REFUSE TRANSPORTATION COST FACTORS FOR PROPOSED STEAM GENERATORS

Site	Avg. Direct Haul Distance, mi.	Refuse Transported, tpd	Average Round Trip Mileage*	Ton-Miles
Canal Road Plant:				
District 1 and District 2 south of Denison Avenue	5.6	400	14.0	5,600
Lakeshore Plant:				
District 2, north of Denison Avenue	4.5	200	11.3	2,260
District 3	4.2	231	10.5	2,426
District 4	5.2	286	13.0	3,718
District 5, SW of 140th Street	1.2	173	3.0	519
District 5, NE of 140th Street	1.9	85	4.8	408
Total		1,375		14,931

$$\text{Transportation cost} = \frac{\$0.20 \times 14{,}931}{1{,}375} = \$2.17/\text{ton}$$

*Includes 25% increase of indirect travel patterns.

Conclusions

At the present time, the City of Cleveland is considering proposals for the operation of new landfill sites under arrangements with private contractors. The new approach will include the use of transfer stations to reduce transportation costs. Including an estimated cost of $0.50/ton for hauling to the transfer stations, it appears likely from recent bidding that net disposal costs will increase to about $7.00/ton. In comparison, the summarized costs (Table 5.21) for disposal with energy recovery in refuse-fired boilers can be considered.

TABLE 5.21: ESTIMATED COSTS FOR CLEVELAND'S WASTE-FIRED STEAM GENERATING SYSTEMS

	District Heating Plant (On-Site Version)		Turbo-Electric Facility
Total Capital Cost, 10^6 $	5.331		46.874
Total Annual Cost, 10^6 $	1.455		12.564
	A	B	
Annual Credit, 10^6 $	0.552	0.519	11.849
Refuse Disposal Cost, $/ton	7.72	8.00	2.62
Transportation Cost, $/ton	2.80	2.80	1.88
Total Net Disposal Cost, $/ton	10.52	10.80	4.50

(A) Based on revenues for steam generated.
(B) Based on displaced coal, labor, and maintenance costs for operating existing, conventional steam generator of equivalent capacity.

It is clear that operation of the combustion-fired power plant would provide considerable cost savings to the City of Cleveland. This is not immediately true in the case of the district heating plant. Assuming that the proposed steam plant were operated in connection with transfer stations, total net disposal costs would still be higher (about $9.75 and $10.00 per ton by costing methods A and B, respectively) than for landfilling. This comparison, however, is based on today's fossil fuel and disposal costs.

As both of these increase, the cost-effectiveness of the proposed heating plant will rapidly change. In terms of fuel cost variations, for example, this sensitivity was demonstrated in Figure 4.9. Another factor that would serve to reduce disposal costs would be an increase in steam sales. This would put the plant on a more attractive (Method A) costing basis and the required increase in the refuse rate would result in a more cost-effective system design.

An alternate plan that can be considered is to construct two smaller turboelectric stations. From the cost model developed on the earlier program, this arrangement can be shown to be nonoptimum and to result in very high disposal costs. It has been concluded, therefore, that the system recommended is the proper approach. To trade off nicely the factors that influence an area case study of this type, it is virtually impossible to realize a perfect balance of system cost benefits in the face of local constraints.

Pyrolysis Processes

Municipal waste is over 80% carbonaceous matter and over 60% cellulosic material as shown in Table 6.1. Through the process of pyrolysis (chemical decomposition by heat in the absence of oxygen) the carbonaceous material can be converted into a readily available fuel gas that can be used as a substitute for natural gas. This would help to conserve our depleting natural gas reserves. Another factor to consider is that municipal waste is not depleting with time, as are the fossil fuels, but is actually being generated in larger amounts each year. This approach to the solid waste pollution problem is being investigated under a grant from the Office of Solid Waste Research and Monitoring of the Environmental Protection Agency by the Chemical Engineering Department of West Virginia University.

TABLE 6.1: MUNICIPAL REFUSE COMPONENTS

Component	As Received (wt. %)	Picked-Dried (wt. %)
Paper	48.0	48.7
Plastic; leather; rubber	2.0	2.8
Garbage (food wastes)	16.0	11.1
Grass; tree leaves	9.0	6.9
Wood	2.0	2.1
Textiles	1.0	0.7
Glass, ceramics, stones	14.0	16.6
Metals	8.0	11.1
	100.0	100.0

FLUIDIZED BED PYROLYSIS

The actual process investigated by the Department of Chemical Engineering at West Virginia University is pyrolysis of municipal waste using a fluidized bed gasifier. The fluidized bed is simply a cylinder containing a bed of high silica sand resting on a gas distribution plate. Compressed gas passes through the distribution plate and into the bed suspending each solid sand particle. This bed is fluid in the sense that a rock placed in the bed will sink while a piece of wood will float. If an opening is made in the cylinder wall, the fluidized sand will pour out while the unfluidized sand will not. These phenomena are illustrated in Figure 6.1.

The fluidized bed was first used as a unit operation to gasify solid fuels (brown-coal) by Fritz Winkler in Germany in 1926, and near the end of World War II carbonaceous waste materials were gasified by the Winkler process. Since solid waste is highly carbonaceous, an investigation of a fluidized bed process to gasify solid waste appeared attractive and unique.

FIGURE 6.1: SIMILARITIES BETWEEN FLUIDIZED SAND AND A LIQUID

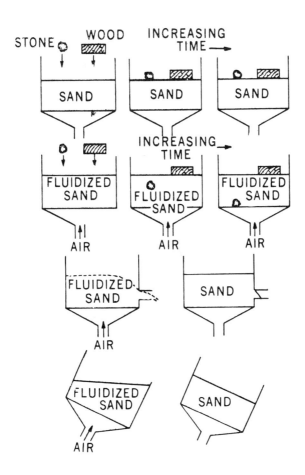

Initial investigations were performed using the fluidized bed as a solid waste incinerator. The fluidized bed operates at practically isothermal conditions and any solids introduced into the bed will immediately attain the temperature of the sand. This characteristic allows for rapid and complete combustion of the solid waste particles. The heat of combustion is immediately carried away by the fluidized sand to be used elsewhere in the bed. The fluidized sand thus acts as a thermal flywheel in that it supplies the heat of reaction to allow the solid waste to burn and then removes the heat of combustion to another area of the bed. It was primarily for this reason that investigations were conducted to study pyrolysis of municipal refuse in fluidized beds.

The process of pyrolyzing the municipal refuse is conducted in a 1500°F. fluidized sand bed in an oxygen-free atmosphere. The cellulose molecule is introduced into the bed as municipal refuse and the thermal flywheel effect immediately brings the refuse to 1500°F. where the cellulose molecule will burn. Due to the absence of oxygen in the bed atmosphere, the molecule instead of burning actually explodes. In this explosion, as in any explosion, the molecule is randomly blown apart. The fragments of the exploded cellulose molecule form methane, carbon dioxide, hydrogen, carbon monoxide, and water molecules. A simple representation of this process is shown in Figure 6.2.

Each cellulose molecule introduced into the fluid bed gasifier undergoes this same chemical decomposition. A by-product of the chemical decomposition is solid activated carbon char which is carried out of the fluidized bed with the pyrolysis gas. The gas and char formed from the chemical decomposition of municipal refuse are of prime interest as sources of energy recycle to our economic system.

If cellulose is burned in an oxygen rich atmosphere, as is the case in incineration, carbon dioxide gas is produced. This is the lowest or most regressive ecological energy level. In other words, this gas can make no positive contribution to our energy resources, and can only be regenerated by the carbon cycle of plant growth and fossilization. The carbon cycle is shown in Figure 6.3. However, when the cellulose molecule is "burned" in an oxygen-free atmosphere the produced pyrolysis gas is, in reference to ultimate resources conservation, ecologically less regressive than the combustion products from incineration. Pyrolysis converts municipal waste to products with a potentially positive energy contribution rather than to carbon dioxide.

The pyrolysis gas can be thought of as water near the top of a waterfall — energy can be produced in turning a paddle wheel. The products of combustion are represented by the water at the bottom of the falls — no possible energy production value. The only possible way to derive energy from the water at the bottom of the falls is to pump it (the carbon cycle) back to the top of the falls. Efficient pyrolysis of municipal refuse in the fluidized bed gasifier allows recycle of a potentially important source of energy and raw materials to our environment as well as reducing overall pollution of land, air, and water.

Process Development

The conversion of fossil fuels into fuel gas or pipeline gas is presently accomplished by complex processes. These processes operate at high temperatures and pressures, require large numbers of skilled personnel, and demand large

FIGURE 6.2: CHEMICAL REDUCTION OF THE CELLULOSE MOLECULE

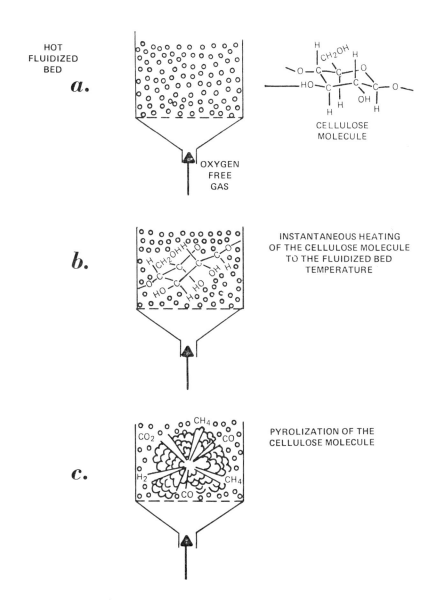

Figures 6.2a and 6.2b show the cellulose molecule being introduced into the hot fluidized bed. Figure 6.2c shows the cellulose molecule literally being rearranged by exploding into its pyrolysis gas products.

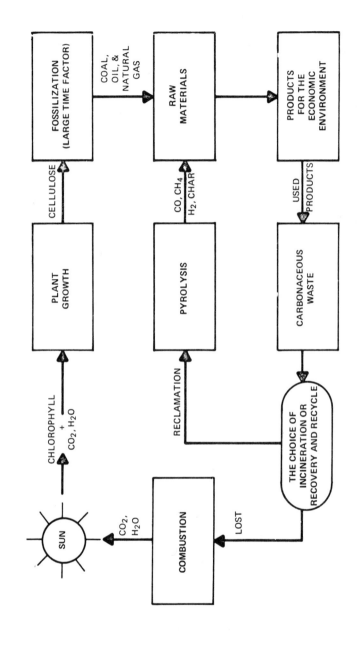

FIGURE 6.3: USING PYROLYSIS TO REINTRODUCE WASTE LOSS INTO THE ENERGY CYCLE

Pyrolysis Processes 111

capital investments. Two such processes which are considered economical to operate in Europe and Asia are the Lurgi process and the Koppers Totzek process. The Lurgi process requires high pressure, pure oxygen, and steam to gasify coal. The Koppers Totzek process operates at 900°C. with moderate pressure, and requires steam and 98.5% pure oxygen to gasify lignite. Both of these processes require a steam generation plant, as well as an oxygen producing facility.

It is easy to see why the capital and operating costs are high, yet it is economical for these plants to produce fuel gas. The gas produced from these processes is not equivalent to natural gas, but it is compressed and pumped as far as 200 miles in pipelines for industrial and domestic use. The fuel gas produced has a heating value of less than half that of natural gas. The economical success of these processes lies chiefly in the fact that Europe and Asia do not possess large supplies of natural gas.

It is clear that the types of processes employed in fossil fuel gasification are too complex for the pyrolysis of municipal refuse. A new attitude must be adopted; process complexity and capital costs must be minimized and additional supporting systems must be avoided.

The factors affecting pyrolysis must also be considered in developing the municipal refuse gasification process. Literature concerning pyrolysis of solid waste is limited, but cellulose pyrolysis has been studied. Since cellulose is the major component of solid waste, an understanding of its thermal decomposition would aid in understanding and predicting what to expect from the pyrolysis of municipal refuse. Some of the major points to consider in the pyrolysis of cellulose and solid waste are summarized below.

> Thermal degradation of cellulose is a complicated phenomenon and several theories have been presented to explain the observed reactions. Figure 6.4 represents the general reactions that are considered to be involved in the pyrolysis and the combustion of cellulose. The data and results from various experiments are scattered and any inferences drawn from these are considered possibilities rather than established facts. The relative proportion of flammable gases produced varies according to the temperature, time, sample geometry, and the environment of the pyrolyzing cellulose. However, certain general observations can be made.
>
> In slow heating, decomposition proceeds according to an orderly arrangement forming increasingly more stable molecules, richer in carbon, and converging toward graphite carbon. In very rapid heating, the molecules are literally torn apart into volatile fractions with almost no possibility of orderly arrangement. Slow pyrolysis at low temperatures yields char and oxygenated gases, and fast pyrolysis yields flammable gases.
>
> In pyrolysis of solid waste, it was found that the rate of heating was an important consideration, that the yield of fuel gas increased with the heating rate, and that flash pyrolysis would result in a significantly higher yield of energy in the gas phase. This is in agreement with the results discussed above for the pyrolysis of

cellulose. The experimental technique used did not allow for really rapid pyrolysis, and it would be assumed that by proper choice of the process, higher yields would be attainable. The temperature used was 1500°F. and it was also noted that the char produced could be further gasified to increase the gas yield.

FIGURE 6.4: REACTIONS FOR PYROLYSIS AND COMBUSTION OF CELLULOSE

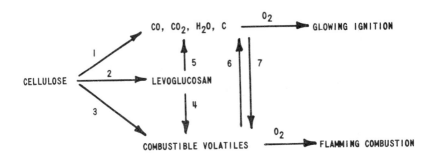

In developing this process for conversion of municipal refuse to a usable energy source for recycle to our economic system, the following items were considered:

(1) Capital costs and operating expenses are to be minimized.
(2) High pressure systems are to be avoided (atmospheric is desirable).
(3) The use of pure oxygen and steam as raw materials is not to be considered.
(4) The number of skilled operating laborers is to be minimal.
(5) The process is to be flexible in size so as to accommodate the municipal refuse output of the particular area that the facility serves.
(6) The process need not generate the equivalent of natural gas to be considered successful. The gas generated must be of sufficiently high value to be compressed and transported a limited distance economically.
(7) Very rapid heat transfer to each solid waste particle to be pyrolyzed with isothermal operation is desirable.
(8) High temperature heating capabilities should exist.

Process Description

Pyrolysis of municipal refuse is an endothermic process. It becomes necessary to rapidly heat the refuse to a high temperature and then add enough extra energy to allow chemical decomposition to occur. This process must be accomplished in the absence of oxygen. The process of combustion requires the same treatment in the presence of oxygen. The main difference is due to the fact

Pyrolysis Processes

that the combustion reactor is exothermic and the ensuing heat must be effectively removed. It would appear that an efficient fluidized bed pyrolysis process could derive the heat required for chemical decomposition from a fluidized bed combustion unit.

The fluidized bed system being developed by the Chemical Engineering Department at West Virginia University for the Office of Solid Waste Research and Monitoring of the Environmental Protection Agency uses the heat given off by the combustion of pyrolysis char to supply the energy needed in the municipal refuse pyrolysis reaction. The oxygen required for combustion is supplied by compressed air, and in order to prevent the nitrogen in the air from diluting the pyrolysis gas, the two reactions are carried out in separate reaction vessels. The transfer of energy between the two reactors is accomplished in a manner similar to that developed by the petroleum industry in manufacturing high octane gasoline.

Each vessel contains equal depths of fluidized sand particles, and the sand can be induced to flow from one vessel to another. The sand flow from the combustion reactor at 1750°F. to the pyrolysis reactor at 1500°F. supplies the heat necessary for the chemical decomposition of municipal refuse to occur. The solid feed to the pyrolysis unit is municipal refuse, while that to the combustion unit is the solid char formed from the municipal refuse pyrolysis reaction. The high heat transfer rate and isothermal conditions of the fluidized bed are very desirable for fuel gas production. A simple schematic of this process is shown in Figure 6.5.

FIGURE 6.5: MUNICIPAL REFUSE PYROLYSIS PROCESS USING FLUIDIZED SAND AND CHAR RECYCLES

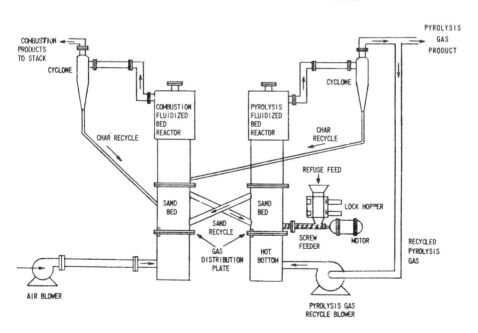

Experimental Gas Production Results

The following is a short description of the experimental results obtained from the fluidized bed pyrolysis of municipal refuse. The tests were conducted using the atmospheric pilot plant fluidized bed gasifier equipment (Figure 6.6) developed at the Chemical Engineering Department at West Virginia University with the cooperation of the Office of Solid Waste Research and Monitoring. The results are presented in Table 6.2.

TABLE 6.2: PYROLYSIS OF MUNICIPAL REFUSE TEST RESULTS

Component	Proximate Municipal Refuse Analysis	Proximate Activated Char Analysis	Gas Production scf/lb. Solid Fed (Dry Basis)	Dry Gas Composition Vol. %	Dry Gas Composition CO_2 Free
CO_2			1.40	16.3	0.0
CO			3.04	35.5	42.4
CH_4			0.95	11.1	13.3
H_2	3.56	2.95	3.18	37.1	44.3
Carbon	25.15	60.82			
Ash	36.54	11.12			
Heating value, Btu/scf dry				366	437

TABLE 6.3: COMPARISON OF THE KOPPERS TOTZEK AND LURGI PROCESSES TO THE FLUIDIZED BED PYROLYSIS OF SOLID WASTE

	Process		
	Koppers Totzek Lignite	Lurgi Coal	Fluidized Bed Solid Waste
CO_2	14.8	25.6	16.3
O_2	0.4	0.0	0.0
CO	46.2	24.4	35.50
H_2	31.9	37.3	37.10
CH_4	2.5	10.3	11.10
N_2	4.2	1.8	0.0
Unsaturated hydrocarbons	0.0	0.6	–
Heating value, Btu/scf dry	210	312	366

The pyrolysis gas could conceivably be treated through a methanation process to raise the heating value to that of natural gas (see Table 6.4). Table 6.3 presents the characteristics of the fuel gas produced by the Koppers Totzek process, the Lurgi process, and solid waste fluidized bed pyrolysis system. Also, the energy recovery from the solid waste fluidized bed pyrolysis process is about 70% while that for the Koppers Totzek and Lurgi processes is between 60 and 65%.

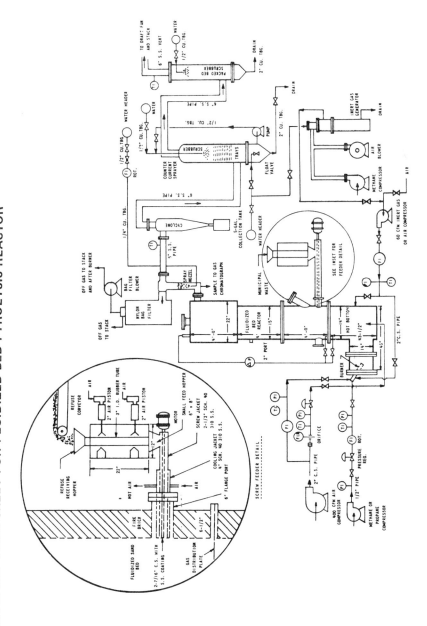

FIGURE 6.6: FLOW SHEET FOR FLUIDIZED BED PYROLYSIS REACTOR

Examining Table 6.3 clearly shows that the gas produced by the fluidized bed pyrolysis process is superior to that produced by either of the other processes. The heating value of the gas produced by the fluidization method is considerably higher than either of the other processes. The Koppers Totzek and Lurgi processes require high pressure, coal or lignite, steam, and pure oxygen. The fluidized bed process uses atmospheric pressure, air, and municipal refuse. Refuse pyrolysis produces better gas, conserves the fossil fuel and natural gas resources, and helps solve the solid waste pollution problem.

400 Ton per Day Facility

The following is a short description of the equipment used to process 400 tons per day of municipal refuse which has an average moisture content of 30%. This description is not meant to be a detailed design layout but merely gives an estimate of the major equipment sizes along with the expected pyrolysis gas production. The schematic of the entire municipal refuse fluidized bed pyrolysis system is shown in Figure 6.7.

The combustion unit is a fluidized bed 3.5 feet in diameter and 20 feet high. The sand bed height is 4 feet with a harmonic mean particle diameter of 0.025 inches. The combustor will be fed the recycled char produced in the gasifier unit at a rate of 31,500 lbs./hr. and will operate at 1750°F. This bed will operate at three times the minimum fluidization velocity using 2.1 million scf/day of air. The off-gas from the combustor will pass through two cyclones to effect gas clean up. The first cyclone will remove large solid particles while the second will remove the smaller particles. The char removed from the cyclones will be returned to the combustor for fuel. The combustion products will then be passed through a heat exchanger to preheat the air entering the combustor for heat conservation.

The fluidized pyrolysis unit will be 7.5 feet in diameter with an overall height of 20 feet. The bed height and sand harmonic mean diameter are the same as for the combustor. The pyrolysis unit will operate at 1500°F. and will gasify 400 tons per day of municipal refuse. The gas to fluidize the pyrolysis unit will be supplied by recycling the pyrolysis gas. One-third of the gas produced will be recycled, and the bed will operate at three times the minimum fluidization velocity. The pyrolysis gas stream will pass into a cyclone to remove the product of activated carbon char produced in the pyrolysis reaction. This char is fed to the combustion unit to supply the heat necessary to keep the fluidized sand temperature at 1750°F.

The off-gas may be passed through an optional processing system. The gas could go through a water gas shift reactor, a carbon dioxide scrubber, a clean up washer; and finally, a methanator to convert all the pyrolysis gas to methane. This system is purely optional as the gas coming directly from the pyrolysis unit is an immediately usable energy source. The energy required to maintain the pyrolysis unit at 1500°F. is obtained from the sand circulating from the combustion unit at 1750°F. The sand circulation rate is approximately 54,000 pounds of sand per hour.

The refuse feed system bears some mention. Most of the system has already been developed for Combustion Power Company's CPU-400 system. The system consists of a refuse storage pit from which the refuse is removed as needed and

Pyrolysis Processes 117

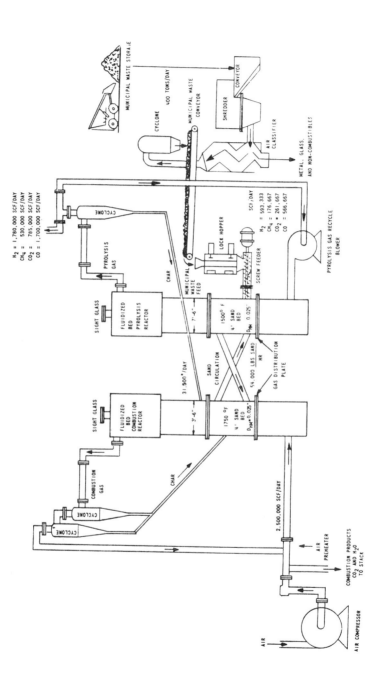

FIGURE 6.7: OVERALL SCHEMATIC OF 400 TON PER DAY MUNICIPAL WASTE PYROLYSIS FACILITY

fed to the conveyor belt by a mechanical lift. The refuse is passed to a sophisticated refuse shredder where the municipal refuse is reduced in size. The refuse is passed through an air classifier where 90% of the metal, glass, and heavy objects are removed. The classified refuse is then fed by conveyor to a lock hopper-screw feeder apparatus where the refuse is fed directly into the fluidized bed. The amounts of pyrolysis gas at different steps in the final gas processing are shown in Table 6.4 on a dry gas basis.

TABLE 6.4: PYROLYSIS GAS PRODUCED FROM 400 TONS PER DAY OF MUNICIPAL REFUSE*

Component	Pyrolyzer Exit scf/Day (Dry Basis)	CO-Shift Exit scf/Day (Dry Basis)	CO_2 Scrubber Exit scf/Day (Dry Basis)	Methanator Exit scf/Day (Dry Basis)
CO_2	785,000	1,610,000	–	–
CO	1,700,000	870,000	870,000	–
CH_4	530,000	530,000	530,000	1,400,000
H_2	1,780,000	2,610,000	2,610,000	–
Total	4,795,000	5,620,000	4,010,000	1,400,000

*Municipal refuse contains on the average 30% moisture.

Four hundred tons per day of municipal refuse would be produced by a city of approximately 100,000 people. The average daily residential consumption of natural gas per capita is 150 scf (dry). It can be seen from Table 6.2 that on a per capita basis 23 scf/day of equivalent natural gas is produced from the pyrolysis unit. Thus, if all the municipal refuse produced per person per day were to be pyrolyzed by the fluidized bed method approximately 15% of the daily natural gas requirement per person could be supplied. This amount of natural gas generation, while it does not flood the market with extra gas, does substantially subsidize the daily natural gas requirements of the United States. Thus, we are able to help solve the solid waste pollution problem, conserve our fossil fuel natural resources, and reduce potential air and water pollution!

Utilization of Pyrolysis Gas

The actual pyrolysis gas from the fluidized bed has many direct applications including direct production of methane as previously described. This gas can be used in residential areas for heating and cooking, and for heating in steel mills. Other direct uses of this gas are for saline water conversion and steam production. Also, the gas could be burned in a jet engine type turbine for power generation. The pyrolysis gas could be used as a raw material in organic industries as well as in mineral and metallurgical industries.

The activated char produced by pyrolysis is also a valuable product. This char can be used directly as a solid fuel to perform some of the tasks of the pyrolysis gas mentioned above. The char can be used for general purification and reclamation of liquid and gas streams. In particular, the char could be used to purify sewage sludge to obtain pure water and then the solids could be used as the energy source for the fluidized bed combustion unit. The activated char could be used to adsorb metallic ions. The char could also be used as the fuel for the fluidized bed combustion unit and could be circulated with the sand. These processes are shown schematically in Figure 6.8.

FIGURE 6.8: POSSIBLE USES OF GAS AND CHAR PRODUCTS FROM THE PYROLYSIS OF MUNICIPAL REFUSE

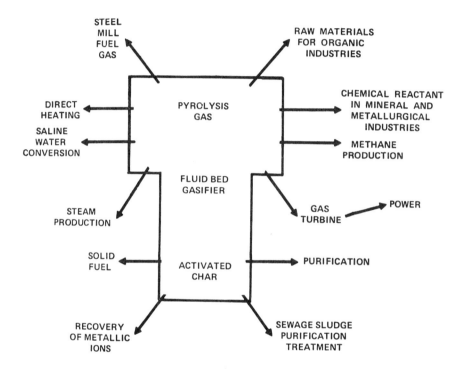

Status

The experimental program carried out at West Virginia University has established that municipal refuse can be converted into fuel gas with a relatively high heating value. It has been demonstrated that pyrolysis in a fluidized bed results in high gas yields with high thermal efficiency. The fluidized test facility used in the experimental program allowed for the feeding of crushed municipal refuse into a controlled atmosphere not unlike those anticipated for a full scale unit.

The work completed has established the viability of the process concept and allowed for evaluation of some of the environmental implications. Preliminary economic evaluations show the fluidized bed pyrolysis conversion process would not increase the cost of disposal but would actually allow for a significant return on the capital investment. It would be premature to give firm cost figures at this time. Before such values are presented a detailed engineering and economic evaluation of a full-scale facility, along with a detailed review of the effects on the total environment, must be completed. Further experimental effort is required before such a detailed evaluation can be completed.

OTHER PYROLYSIS PROCESSES

The information in this section is based upon a 1973 report prepared by General Electric in cooperation with the Connecticut Department of Environmental Protection.

Pyrolysis can generally be classified as the thermal breakdown of material in the absence of oxygen, prevalent in coke and charcoal ovens. Recent developments of pyrolysis of solid waste include processes to produce an oil (the Garrett process) and to produce various forms of low Btu fuel gases (Union Carbide Corporation and the Urban Research and Development Corporation).

Other pyrolysis processes are intended to serve as high-volume-reduction, low-pollution incinerators with steam heat recovery, (Monsanto-Landgard and Torrax), but can also be used to generate low Btu fuel gas. Figure 6.9 illustrates the Monsanto-Landgard Process.

The Garrett Process (Figure 6.10) utilizes an extensive amount of front end preparation to separate the organic material from the inorganics. The organic material is very finely ground in the fine grinder to -20 mesh and then fed to the pyrolysis reactor which produces a gas and a char. The gas is condensed to produce a fuel oil somewhat comparable to the No. 6 fuel oil used in power plants. Garrett also recovers glass and ferrous metal from the inorganic stream; and the process also offers the potential for extracting aluminum.

The process has been demonstrated in a small-scale pilot plant, and EPA has recently issued a Demonstration Grant to build a larger scale facility in San Diego. The oil produced will be burned in a San Diego Gas and Electric Company steam power plant. Garrett has also demonstrated the capability to produce a fuel gas by operating the pyrolysis unit at higher temperatures. However, they feel the oil process is more economical.

The Union Carbide, U.R.D.C., Torrax and American Thermogen processes, on the other hand, require no front end shredding or separation of waste. The only limitation is that bulky items be cut up small enough to go in the charging doors. Each of these four uses a vertical shaft furnace, similar in appearance to a blast furnace. Each is a slagging process where local temperatures are high enough to melt everything, producing an inert slag occupying from 2 to 6% of the volume of the incoming waste. Water quenching the slag breaks it up into a frit about the size of coarse sand. This material has potential use as an aggregate for construction and road building applications.

The Union Carbide process (Figure 6.11) is unique in that it employs nearly pure oxygen in its process. An on-site oxygen plant feeds oxygen into the bottom of the shaft furnace to supply the high temperature combustion zone. The refuse is fed in through an air lock at the stop. Only enough oxygen is fed to the shaft furnace to allow burning in the combustion zone near the bottom of the furnace. This provides enough heat to melt the noncombustibles at the

FIGURE 6.9: FLOW CHART OF MONSANTO-LANDGARD SYSTEM

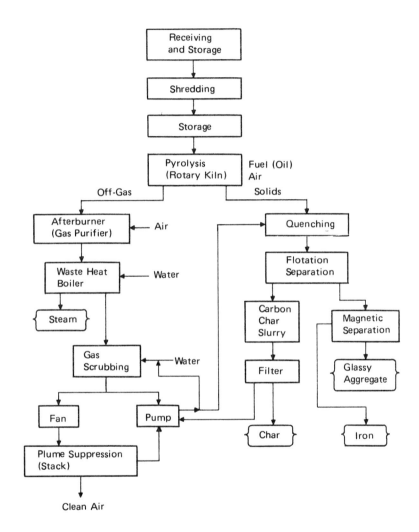

FIGURE 6.10: FLOW CHART OF GARRETT PROCESS

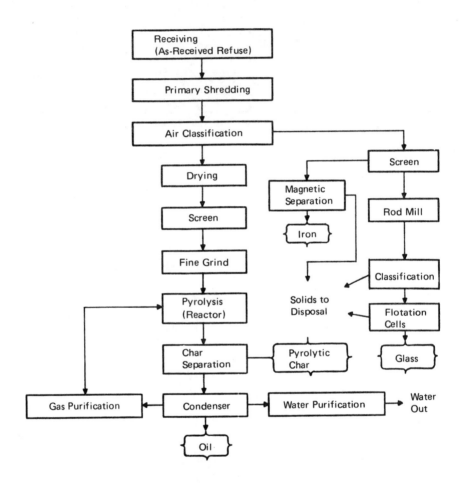

FIGURE 6.11: FLOW CHART OF UNION CARBIDE SYSTEM

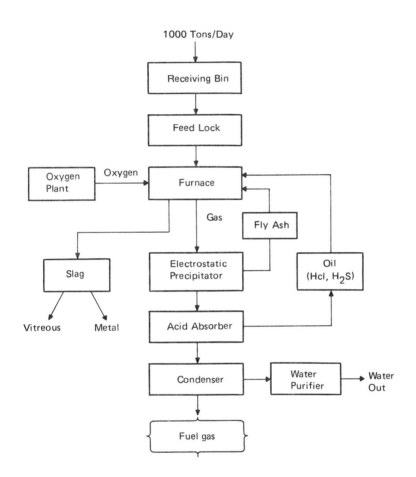

bottom and allow the melt to be tapped off as a slag. Above the combustion zone, the oxygen has been exhausted and combustion of the incoming refuse at the upper end of the shaft furnace cannot be supported.

The heat from below dries the incoming refuse and volatilizes it to produce a fuel gas with a heating value of approximately 300 Btu per standard cubic foot (natural gas has a heating value of about 1,000 Btu per standard cubic foot). This gas is then cleaned up to remove acids, moisture and particulate matter. The gas has the potential to be upgraded to pipeline quality gas by further processing.

The U.R.D.C. process uses preheated air instead of nearly pure oxygen to supply the combustion zone of its shaft furnace. It produces a lower quality gas (estimated to have 100 to 160 Btu per standard cubic foot heating value), but potentially in greater quantity. The U.R.D.C. product gas has not yet been fully evaluated.

> Various pyrolysis processes are described in
> the final chapter of this book.

European Practice

OVERVIEW

Incineration practice in Germany, and in all Europe, is based on an entirely different set of conditions than in America. In many German refuse plants either heating steam or electricity from steam is produced from the heat of the refuse burning. The German decision to produce steam from refuse stems from a compound line of reasoning and conditions. Although German refuse has a lower calorific value than American refuse, the difference in calorific value is less than the difference in fuel costs between the two countries.

Thus, the economic potential for energy production from refuse is somewhat more favorable in Germany than in the United States. This alone, however, would not be sufficient to justify all the refuse incinerators with power production facilities. Additionally, because West Germany is so densely populated (8 times the population density of the United States), a high degree of environment management is a necessity. High-quality stack emissions are paramount. To do a good job of cleaning combustion gas, its temperature must be reduced, usually to less than 600°F. One good way of cooling gases is by heat transfer, to absorb the heat by producing steam.

There is also a third factor, which is somewhat more intangible. Because power and utility services are performed by local governments, conservation and waste conversion, the production of energy from municipal refuse, can be more easily practiced, as they are less dependent on profit motivation. The actual practice of incinerating refuse with steam production takes two basic forms:

1. The primary operation is refuse burning, steam production is incidental, and the quantity of steam so produced and the time of production are not tailored to the community's steam or electric power needs. Rosenheim, Berlin and Frankfurt operate on this principle. Conventionally fueled boilers must be available to meet the maximum heating or power demand of the community. The refuse boiler merely reduces the quantity of conventional fuel burned.

2. The incinerator is operated primarily to produce steam or electricity in the amount and at the time it is needed, refuse is burned to this schedule, with auxiliary fuel being used to supplement the fuel need. This is the principle followed at Munich and Dusseldorf. Here again, conventionally fueled equipment nearly equal to the maximum steam or power demand must be available, but the fluctuation of the load on this conventionally fueled equipment is much reduced. It was pointed out that there are valid arguments for both systems, and the choice must be tailored to the individual city.

Other factors include whether the refuse incineration will occur in a separate firebox and boiler, and whether refuse burning and conventional fuel burning will be used with one set of boiler tubes. In addition, it must be decided whether the refuse incinerator will be a simple unit producing low pressure steam (20 to 50 psi) or will contain a more efficient but more complex and costly high pressure (400 to 1,300 psi) boiler. The low pressure steam can be used directly for municipal heating, or can be fed into a high pressure, conventionally fired boiler.

Steam and power generation from municipal refuse is not a profit making activity in Germany. Refuse is not a free source of fuel, because it costs more in equipment, controls and manpower to burn refuse than it does to burn conventional fuels (lignite, coal, oil or gas). The cost of producing a ton of steam or a kilowatt hour of electricity from refuse is often expressed in terms of how much more it costs with refuse than with conventional fuel; the extra cost is chargeable to waste disposal for it would cost that amount to burn it without steam generation or to bury or compost it.

Solid waste management in Germany is an impressive operation. The West German federal government and the public works departments of the various cities are doing an excellent job in this difficult area. The public works departments and the officials within these departments have a great deal of initiative and appear to have a relatively high degree of independence. The invention and testing of the Dusseldorf grate is one example; the ash-sintering design at Berlin another; even the setting of the waste disposal fee, as at Frankfurt, is another.

The municipal officials, to whom the public works officials report, appear to have a generous and approving attitude toward waste management costs. Such factors as the architectural appearance of facilities and the safety and welfare of the employees are favorably weighed. Engineering is not done on an absolute minimum cost basis, but rather on an optimum design basis.

DESCRIPTION OF SPECIFIC GERMAN PLANTS
Munich North, Block I

This plant consists of two identical Benson-type units. These are the oldest of the units under consideration and are characterized by twin chamber furnaces; i.e., the refuse and coal furnace chambers are separate but share a common tube wall. The combustion gases are combined at the top of the furnace chambers, and pass through a common superheater and economizer. All of these elements comprise one furnace setting or unit. Each unit includes a Martin (backward reciprocating) grate for municipal refuse combustion and a suspension-fired furnace chamber for the combustion of pulverized coal.

Steam conditions for each Block I unit are 220,000 lbs./hr. of 2,600 psig steam at 1004°F., while firing 660 tpd refuse plus auxiliary coal. Maximum continuous load is 220,500 lbs./hr. of superheated steam at 2,650 psig and 1004°F. The reheat steam flow at this load is 198,000 lbs./hr. at a pressure of 1,180 psig and 1004°F. Ferrous metals are removed from the combustion water-quenched residue by magnetic equipment.

Munich North, Block II

This unit was designed in 1966. It was evolved from the Block I units, but with one important design change. The Block II unit is a single-chamber furnace, with pulverized coal combustion occurring directly above the refuse grate. Steam quality is identical to that of the Block I units; steam production at 800,000 lbs./hr., is considerably higher. All the electrostatic precipitators of the Munich North plants are of Lurgi (Frankfurt) design and are horizontal-flow, steel shell precipitators having pyramidal hoppers. The characteristic differences between the Munich North plants can be seen from the following summarization.

Comparative Information on Munich North Plants

	Block I	Block II
No. of turbines	1	1
No. of steam generators	2	1
Refuse heat input, % (LHV)	40	20
Refuse rate, tpd	660	1,060

Dusseldorf

This plant consists of four essentially identical boilers, arranged in pairs. The Dusseldorf furnace is primarily for firing refuse, although there are auxiliary oil guns that can be used for start up and when the heating value of refuse is low. The refuse is fired on a roller grate; as at Stuttgart, only bulky refuse is shredded. The combustion air can be directed over a steam air preheater and a feedwater/air preheater if heating of this air is desired. Three electrostatic precipitators treat the combined flue gases of four units. There is also provision for recirculating waste gas. Each steam generator is designed to deliver from 25,500 to 35,200 lbs./hr. of steam at 1,280 psig and 932°F. The roller grate of VKW design is designed to burn 22,050 lbs./hr. of refuse with an exit gas temperature of 410°F.

Stuttgart

The Stuttgart plant consists of two units, which are nearly identical. Both units have one oil furnace and one refuse furnace with the gases combining before entering the convection section. As with the other German units considered, there is provision for recirculation of the flue gases to cool the residue. The units have steam/air and waste gas/air (panel design) air heaters. The steam generators are designed to deliver 204,600 lbs./hr. steam at 925 psig and 977°F. for normal operation with either oil-firing or combined-firing. The maximum continuous power level is 275,600 lbs./hr. steam at the same conditions. The boilers were designed to handle 40,920 lbs./hr. of refuse having a lower heating value of 2,159 Btu/lb. The refuse furnace volumes of Units No. 28 and 29 are 17,655 and 17,443 cu. ft., respectively. The oil furnace volume is 13,277 cu. ft. in both units.

The noted difference between the two Stuttgart units is the grate designs. Unit 28 is equipped with a Martin grate, while Unit 29 is equipped with a roller grate that evolved from the Dusseldorf (VKW) design. Only bulky refuse is shredded before burning. Ferrous metals are removed from the residue magnetically.

It is interesting to note that the fly ash emissions for the two Stuttgart boilers were expected to be identical, at 1.81 gr./scf. The Martin grate furnace, Unit 28, closely approached this figure during Technischer Uberwachungs-Verein (TUV) testing, but the roller grate unit (Unit 29) emitted approximately 25% less fly ash under similar test conditions. The grate areas are very similar, but the Unit 29 underfire air is approximately 35% lower than Unit 28. Later published test data showed Unit 29 to be producing about 30% more flue gas particulates than Unit 28. Such variations must be expected, considering the nature of the fuel. The two Stuttgart units are each equipped with one electrostatic precipitator of Rohtemuhle design, similar to the aforementioned Lurgi units.

Data Analysis

In accordance with standard German practice, the formal contract acceptance tests for each of these plants were performed by the TUV. The TUV is a state sanctioned agency that reviews and approves final design and performs acceptance tests on virtually all publicly owned capital facilities. Transcripts of TUV acceptance test data on the above plants were procured and reviewed.

A primary objective was to be able to predict, in quantitative terms, the nature of the emission problem that could be expected to result from the application of refuse-fired or combined-fired system elements delineated by the study. This is obviously necessary in order to specify the required control techniques for such a system. Secondly, the reviews of German combined-fired practice could form the basis for selecting with confidence the required control techniques. Finally, industrial experience must be brought to bear in the course of designing and cost-estimating the control systems required for each of the study's output system recommendations.

Table 7.1 is an overall tabulation of the characteristics of the plants under consideration. Table 7.2 is a detailed tabulation of the precipitator design data for these plants.

OTHER REFUSE-FIRED PLANTS
Essen-Karnap

This plant has a Lindemann Shear and also a refuse-burning traveling grate. It was originally designed for firing pulverized coal and modified to burn, in addition, sewage sludge and refuse. Perhaps the most interesting aspects of this

TABLE 7.1: GERMAN PLANT DESIGN DATA

	Munich North		Dusseldorf	Stuttgart	
	Block I (2 Units)	Block II	4 Units	Unit 28	Unit 29
Furnace type	Combined-fired twin chamber	Combined-fired single chamber	Refuse only	Combined-fired twin chamber	Combined-fired twin chamber
Date commissioned	1962	1966	1965	1965	1965
Refuse grate					
Type	Recip./backward feed	Recip./backward feed	Roller or drum	Recip./backward feed	Roller or drum
Manufacturer	Martin	Martin	VKW	Martin	VKW
Area, $ft.^2$	605	1,035	275	543	550
Charging rate, $lb./ft.^2$-hr.	91	87	76	81	–
Btu release, $Btu/ft.^2$-hr. (LHV)	455,000	435,000	378,000	410,000	404,000
Under-fire air, SCFM	–	–	Flue gas recirc.	34,300 at full load	22,000 at full load
Refuse rate					
Short tons/day	660	1,060	250	492	530
lb./hr.	55,000	88,500	20,800	41,000	44,300
Aux. fuel	Coal	Coal	None	Oil	Oil
Steam condition					
Production 10^3 lb./hr.	220	800	32	205	205
Pressure, psig	2,600	2,600	1,280	925	925
Temp., °F. (SH/RH)	1004/1004	1004/1004	932	977/–	977/–

(continued)

TABLE 7.1: (continued)

	Munich North		Dusseldorf	Stuttgart	
	Block I (2 Units)	Block II	4 Units	Unit 28	Unit 29
APC equipment					
Type	Elect. pptr.	Elect. pptr.	Elect. pptr.	Elect. pptr.	Elect. pptr.
Manufacturer	Lurgi	Lurgi	Lurgi	Rothemühle	Rothemühle
Rated flow	–	Various	Various	172,000 acfm	172,000 acfm
Collection efficiency	99.53%	99+%	99+%	98%	98%

TABLE 7.2: ELECTROSTATIC PRECIPITATOR DESIGN DATA FOR GERMAN PLANTS

	Munich North		Dusseldorf	Stuttgart	
	Block I	Block II	4 Units	Unit 28	Unit 29
Number of boilers/pptr.	1	1	2	1	1
Number of ducts	34	84	28	42	42
Duct width, in.	9.5	–	8.5	8.75	8.75
Duct height, ft.	24.6	27.4	20.6	25.4	25.4
Duct length, ft.	29.1	31.5	18.9	16.4	16.4
Total proj. coll. area, ft.^{2}	48,700	–	21,800	35,000	35,000
Inlet cross-sect. area, ft.^{2}	658	1,810	406	780	780
Transformer-rectifier sets	Two-650 ma.	Two	Two	One 500 ma.	One 500 ma.
Operating voltage, kv. DC (max.)	76	–	–	–	–
Number bus sections	2/series	2/series × 2/parallel	2/series	2/series	2/series
Design gas velocity (max.). ft./sec.	3.4	3.16	3.7	3.67	3.67

plant are that it is a total waste facility and that it is one of the few plants built and operated by a private utility, the Rheinisch-Westfalisches Elektrizetatswerk AG (RWE).

In this area all sewage is delivered to the Emse River (parallel to the Ruhr River), from which open system it is later withdrawn for treatment. Purified water is finally released to the Ruhr River. The clarification sludge is brought to the plant on a long conveyor belt. An interesting feature of the continuous belt is that it is twisted at each end so that the belt rollers located below the belt are always in contact with the unused side of the belt.

Essentially three types of refuse are delivered to the plant. Municipal refuse is brought in by regular municipal trucks. Bulky refuse is also delivered, usually by private vehicles, and is processed by a Lindemann Shear. Industrial chemical refuse is accepted in a special pit or in liquid storage tanks. Reduced bulky refuse is discharged to the same pit where municipal refuse is dumped. There are doors at each truck stall which open just far enough for the truck to discharge. An inclined apron is provided so that the refuse crane cannot possibly hit a truck or door. The pit, which is some distance from the refuse-burning facility, is designed to be under negative pressure, but the odor near the pit-building is quite noticeable.

A feature of this plant is that the raw refuse is sent to a magnetic separation-step prior to delivery to the furnace. The recovered metals are collected, baled and removed daily. The metal is used in a foundry producing cast iron; the small amount of tin present in this scrap is apparently considered acceptable for this iron. This certainly does not apply in the case of steel production. It was pointed out that metal baled after combustion contains a great deal of ash which has to be removed before using the metal in a melt. For this reason, the scrap commanded 50% premium over burned scrap.

From the metal separation step, the refuse is delivered by conveyor belt to the boiler house, where there are ten steam generators delivering steam to five steam turbines. Five of the steam generators have been modified to burn refuse. The other five are fired on sludge. Boiler 3 was modified in 1961, boilers 1 and 2 in 1969 and boilers 6 and 7 in 1964; total capital costs for the refuse-handling modifications amounted to $6.2 million. Both the clarification-sludge and the refuse are brought to the respective furnaces by conveyor belt; 2,000 tons of refuse and a like quantity of sludge are handled daily.

Each of the five refuse furnaces is completely refractory enclosed, with no heat-absorbing surface. The flue gas of the combusted refuse is vented to the water-cooled boiler at a point below the tangential coal burners. Average heating value of the refuse is 2,160 Btu/lb. (LHV); the highest noted was 2,880 Btu/lb. (LHV). It was confirmed that this plant had never experienced any corrosion of tube surfaces in any of the units. The only tube wastage attributable to refuse-firing resulted when occasionally the refractory furnace was overloaded and the flame impinged on the waterwalls of the steam generator.

Recovery of ash, both bottom residue and fly ash, is well handled. German regulations do not permit the use of fly ash for making concrete, while Dutch specifications are not restrictive. Almost all of the fly ash collected (1,000 tpd) is sold to Dutch distributors who haul it to Holland in their own trucks where they resell it at a profit.

With regard to a private utility handling refuse, it was pointed out that the operation was somewhat forced upon them by the local municipality. The RWE agreed to handle refuse in exchange for providing power to some sections of the area previously served by municipal power. It was explained that the RWE is compensated for handling the refuse and sludge in a rather complicated manner, which assures a normal return on the investment.

Berlin-Ruhleben

This plant is rather unique because it is actually three plants in one. It is a refuse-incineration plant, a clinker-processing plant, and a clinker-sintering plant. The completed structures include six boilers. Steam is delivered to the existing Reuter power plant across the river. Superheated steam is delivered at or above 905°F. (940 psig); the steam-flow chart indicates considerable flow variation. The arrangements with the power plant involve two rates of payment. A lower payment is given when the steam temperature falls below 905°F. When this occurs the steam from the refuse-burning plant is diverted from the high pressure turbine to the intermediate pressure turbine. It is claimed that this is automatically controlled.

The plant has suffered some corrosion in the furnace, due to localized reducing atmospheres. An increase in the excess air used has virtually eliminated this corrosion. Along the roller grates in this furnace are waterwall tubes which eventually become part of the side wall. These tubes have also suffered some wastage. This type of construction is also used in Stuttgart and Mannheim but more for the purpose of abrasion protection than for heat pickup.

While some wastage has been noted at Stuttgart, none was reported at Mannheim. Some tubes in the Berlin-Ruhleben plant have failed because of longitudinal cracks. In several places, ash had collected behind the tubes, which thus were being pushed out into the furnace. These units do not have welded walls. There was heavy ash accumulation in the superheater, although no plugging had yet occurred. It was claimed that the refuse in Berlin contains more ash than in most other German cities.

A novel refuse-feeding arrangement, consisting of a continuous, tank-track conveyor, is used between the chute and the furnace grate. Most European plants use table-type feeders. Generally, these feeders are not used in the United States.

The sintering plant had not been as successful as expected. When product from this plant was mixed with concrete, cracks developed in the cured material. This had not been a problem during pilot plant evaluations. The cause is believed to be elemental aluminum. Apparently aluminum foil has been marketed only recently in Berlin. It is claimed that the aluminum passes through the furnace unoxidized and failure of the concrete is caused by the reaction of the free metal with the alkaline cement slurry. After working with Battelle on this problem, it is now believed that by washing the residue in lime solution the aluminum can be dissolved. The sintering plant is operated at least several days each month.

Munich-South

This plant is very clean and impressive. The combined refuse/natural gas facilities are Units No. 5 and No. 6. The other units are coal-fired boilers. In the control room the steam-flow trace is very smooth, exhibiting much less variation than did the units in the Munich-North plants. The grate used in this unit was a Martin design, the refuse furnace was by VKW, and the natural gas-fired steam generator was constructed by Deutsche Babcock. This unit also has a capability for future coal-firing.

The flue gas leaving the refuse-fired economizer is too hot for introduction into the electrostatic precipitator. It is therefore mixed with the cooler flue gas produced by the natural gas-fired boiler. If only refuse is burned, the flue gas will have to be cooled by water sprays.

Only refuse collected by municipal trucks is brought to this plant. Bulky refuse is handled by special trucks equipped with built-in shredders. The importance of having at least that degree of control over refuse size was stressed. The residue is removed by conveyor belt and is first taken to a magnetic separator. Baled scrap is loaded into railroad cars and residue is taken to landfill by truck. Double doors are employed at the refuse pit, with interlocks provided to prevent both doors from opening simultaneously. Air is withdrawn from the top of the pit.

As in Berlin, one cannot smell any refuse unless standing at the edge of the pit. An interesting point on the architecture is that there are several floors of offices located above the unloading dock. The crane operators are located in a pulpit at the top of the pit, similar to the Berlin operation. The pulpit is located at an elevation even with the chute. However, unlike the Munich-North plants and the Essen-Karnap plant, it is physically possible for the crane to hit an unloading truck unless a stop on the bridge of the crane is provided.

Mannheim

The Mannheim plant is located on Friesenheimer Island in the Rhine River; there are some industrial complexes on the islands as well as a landfill. When the refuse-burning plant is not operable, refuse trucks are diverted to the landfill; steam demands are fulfilled by operating standby, oil-fired boilers. At the time the installation was planned, the various grates available were studied but none of them was considered particularly well suited for refuse-burning. The choice of a traveling grate was based simply on the fact that it was the cheapest machine available. It was found to be advantageous, however, to use several grates in order to achieve some agitation by tumbling.

When the plant was first started some corrosion was noted, but it appears to have subsided somewhat. The convection sections of the steam generator, several of which have staggered-tube arrangements, are similar to the Stuttgart unit; it would therefore seem that erosion may have been as much the cause of tube wastage as corrosion.

This unit is similar to the units at Berlin and Stuttgart in the use of several rows of waterwall tubes parallel to the grate for abrasion protection. At Mannheim, however, the feedwater is first sent through these tubes before flowing to the economizer. At Berlin and Stuttgart, the abrasion protection tubes are part of the boiling section. No failure of these abrasion protection tubes has occurred at Mannheim. It would appear likely that tube wastage of this section would be dependent upon metal temperature. The Mannheim plant is equipped with electrostatic precipitators. As in Stuttgart, a Hazemag shredder was used to reduce bulky waste.

Frankfurt am Main

Some corrosion problems have been experienced here, but not of an unmanageable degree. The steam produced here is for district-heating and hot water supply (through heat exchangers) in an adjacent apartment house complex. The steam produced by refuse-burning is an auxiliary source to that from conventional, oil-fired boilers. One of the unique features of the refuse-burning units is that the refuse from the crane is dumped on a vibrating trough, which in turn discharges to a vertical chute. While an interesting feature, it is doubtful that it is a necessary one. No other plants are known to have such an arrangement, and all appear to work well without it.

As in other plants, the pit and building have been built with a view to future capacity requirements. With extra pit capacity, the operators try to stagger refuse deliveries on a weekly basis, so that they maintain some week-old refuse. It is claimed that more uniform burning is achieved by mixing aged and fresh refuse. The furnace volume was found to be extremely generous, and no overfire air was used. Judging from the fact that the excess air was nearly 100% and the flames produced were somewhat lazy and spotty, the grate surface area was probably oversized. The residue from this plant revealed much unburned material, including partially burned paper. As in Munich, trucks with built-in shredders were employed, in lieu of stationary shredding equipment at the plant.

The most persistent maintenance problem has involved the crane cables which had to be replaced every few weeks. The cranes in this plant were equipped with an automatic control system, so that once a bucket was loaded the charge could be automatically taken and discharged at a predetermined chute. However, in the automatic mode the crane could move in only one direction at a time. This resulted in an unacceptably slow feeding rate. Manual operation, in which simultaneous tridirectional control is routinely achievable, therefore, had to be adopted.

Issy-les-Moulineaux

This plant is located near the Seine, just outside of Southwest Paris. It contains four refuse-fired, natural circulation boilers, which utilize auxiliary fuel (oil) only on start-up. The rated plant capacity is 60 tph while operating at steam conditions of 770°F. and 925 psig.

Samples of refuse, amounting to approximately 5 tons, are taken 10 days of the year. It is claimed that samples taken during the summer indicate heating values higher than the steam generator calculated heating value. During the winter, analyzed samples are lower than calculated by records. The furnace exit temperature is maintained below 1000°F. The average LHV is 3,600 Btu/lb. The minimum LHV is 1,600 Btu/lb. The grate metal temperature is between 300° and 400°F. In this plant, as in some other plants, a siftings hopper is located under the table feeder. Uncompressed, recovered metals are sold.

In-line tube spacings are used throughout the convection sections. The air is heated in a steam-coil air heater. The first corrosion noted occurred after 5,000 hours, while superheater trouble occurred after 14,000 hours. Refuse is collected by municipalities or private contractors. The plant receives money from the municipalities at the end of the year and the amount is dependent, in part, on the plant operation.

Ivry

This plant is also located near the Seine, but outside of Paris to the southeast. The designed operating conditions are 875°F. and 1,400 psig, using natural circulation boilers. Like the Issy plant, a reciprocating grate is employed and refuse constitutes the sole fuel used, except on start-up. It was predicted that upon completion, Ivry would be the world's largest refuse-fired steam generation facility, with the two furnaces being able to handle 2,400 tpd. The completed plant cost will be $30 million. As elsewhere, the pit operators are located in a pulpit. However, at Ivry the pulpit is located at a point lower than the chute. The operator has therefore been provided with a closed-circuit television and constantly has a view of the chute on the screen.

EMISSION CONTROL EQUIPMENT

The most impressive and laudatory feature of German refuse incinerators is the quality of the stack emission. Fly ash is removed at the turns in the boiler and flue gas passages, and it appeared to be economically and satisfactorily managed in all plants. The fly ash is conveyed to the burned-out incinerator residue (clinker and ash) in the dry state at most plants, but at Frankfurt it is conveyed in a water slurry and settled out in a separate operation.

Large, heavy-duty electrostatic precipitators designed for 98 to 99% efficiency are incorporated into all the incinerators. These precipitators are the only gas-cleaning equipment used, no prior scrubbing, centrifuging or filtering. The electrostatic precipitators are continuous flow-through, with periodic shakedown self-cleaning.

The German air pollution control standards allow a maximum of only 150 mg. of particulate material per cubic meter of gas cooled to standard condition (760 mm. pressure, 60°F.). This corresponds to about 0.192 lb. of particulates per 1,000 cubic feet of flue gas corrected to 12% carbon dioxide. The present U.S. guideline is 0.428 lb. of particulate per 1,000 cubic feet. The quality of the German exhaust is thus very good. The German refuse incinerators are generally equipped with very high (up to 300 feet) chimneys. This provides excellent dispersion of the gases above the city.

Gaseous Emissions

Compared to the situation in this country, there is less reason for German national concern over control of sulfur oxides pollution. European coal is notably low in sulfur (less than 1%). Fuel oil burned in Europe is also low in sulfur for the most part. Based on the limited data available, derivations of sulfur balances were made. This subject is discussed in the main volume of the report.

Process Descriptions

Various selected processes are described in this chapter, based upon information gathered by Midwest Research Institute in conjunction with a project conducted for the President's Council on Environmental Quality. Please keep in mind that many of these processes are still under development, and the technology is changing rapidly, particularly in view of the rapid rise in the price of fuel oil. Therefore, some of the technical details may have changed since the time the information was originally collected.

HORNER AND SHIFRIN FUEL RECOVERY PROCESS

This process was developed for the demonstration plant in St. Louis, with an input of 650 tons/day (2 shifts) and an output of 600 tons/day of fuel, and 50 tons/day of magnetic metals.

Figure 8.1a shows diagrammatically the elements of the refuse processing facilities. Raw refuse is discharged from packer-type trucks to the floor of the raw refuse receiving conveyor. From the receiving conveyor, the refuse is discharged to a belt conveyor, which in turn discharges to a vibrating conveyor, which feeds the hammermill. The hammermill discharges to a vibrating conveyor, which feeds a belt conveyor leading to a storage bin. (Consideration is being given to adding an air-classifier after milling.) Magnetic separation is effected at the head pulley of this belt conveyor. From the storage bin, the processed material is conveyed to a stationary packer, which loads trailer trucks for shipment to the power plant, about 18 miles from the processing facilities.

Figure 8.1b is a diagram of the facilities at the power plant. The self-unloading mechanisms of the transport trailers discharge the supplementary fuel to a receiving bin, from which the material is conveyed to a pneumatic feeder for transfer to a surge bin. The surge bin is equipped with four drag chain unloading conveyors, each of which feeds a pneumatic feeder. Each of these four pneumatic feeders conveys the supplementary fuel directly to a firing port in each corner of the boiler furnace.

Process Descriptions 137

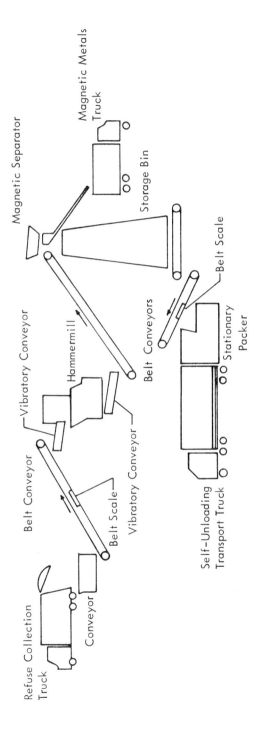

FIGURE 8.1a: HORNER & SHIFRIN FUEL RECOVERY SYSTEM — PROCESSING PLANT

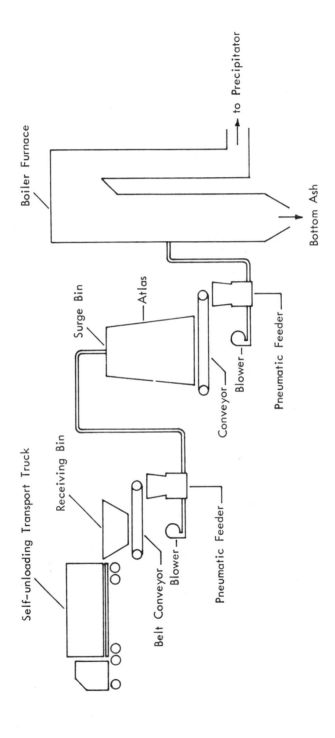

FIGURE 8.1b: HORNER & SHIFRIN FUEL RECOVERY SYSTEM – POWER PLANT FACILITIES

The boiler is small when compared to the newer units in the Union Electric Company system, but it is of modern, reheat design, and the test results from this unit should be applicable to numerous other existing similar units in service in many parts of the country. Built by Combustion Engineering, Inc., who also cooperated fully in the original study, the unit has a nominal rating of 125 megawatts, and will burn about 56.5 tons of Illinois bituminous coal per hour at rated load. The unit is tangentially-fired, with four pulverized coal burners in each corner. It also is fitted to burn natural gas. The furnace is about 28 feet by 38 feet in cross section, with a total inside height of about 100 feet.

At full load, the quantity of refuse equivalent in heating value to 10-20 percent of the coal will be about 12.5-25 tons/hour, or 300-600 tons/day. It is intended to fire the refuse 24 hours/day, but only 5 days/week, since City of St. Louis refuse collections are scheduled on a 5 day/week basis. No difficulty in boiler operation accruing from this interrupted refuse firing schedule is anticipated.

KINNEY THERMAL RECOVERY SYSTEM

A.M. Kinney, Inc. have completed an engineering design for a plant to process 1,000 tons/day of solid waste.

Figure 8.2 presents a schematic diagram of the thermal recovery unit to process 1,000 tons/day of municipal refuse. Only one train of equipment is shown.

Refuse is weighed and recorded at the scale house and then discharged into an enclosed pit or receiving area. Bulky items - engine blocks, bed springs, appliances - are removed for salvage or direct burial. Such items comprise 1% or less of total volume. The remainder is conveyed to a Black-Clawson Hydrapulper which converts all pulpable and friable materials to an aqueous slurry.

Nonpulpable materials are ejected continuously from the Hydrapulper, flushed off, conveyed to a drum washer and thence to a magnetic separator where ferrous material is recovered.

The slurry, 3 to 5% solids, is pumped to a dump chest for blending. Mainly it consists of fibrous organics plus small pieces of glass, metal, ceramics and grit. From this chest it is pumped to a liquid cyclone separator where the nonfibrous materials are removed by the centrifugal action. These materials may be processed further into aluminum and glass fractions for recovery.

The slurry goes from the cyclone to a surge chest for further blending, and then to screw-type thickeners which dewater it to about 10% solids. From here it goes to a high density press which produces a cake with 50% solids content. This cake is broken into 1/2 inch to 1-1/2 inch chucks by a high speed screw conveyor. All water removed is returned to the pulper.

At this point the original mix of garbage and trash has been converted to a nearly homogeneous material. It meets the criteria for power boiler fuel and may be fired with or without additional processing, depending on the type of power boiler used.

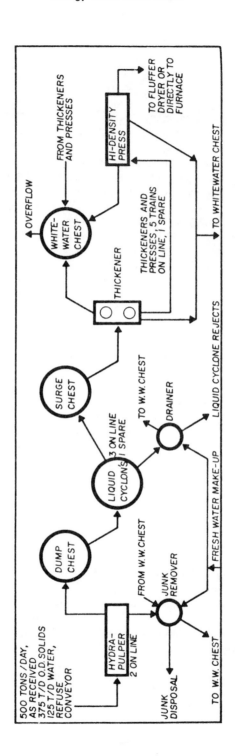

FIGURE 8.2: KINNEY THERMAL RECOVERY SYSTEM

COMBUSTION POWER COMPANY PROCESS

The CPU-400 process is being developed in an 80 ton/day pilot plant for later scale-up to a 1,000 ton/day commercial plant that would generate 17,740 kilowatts of electricity.

The CPU-400 pilot plant is shown in Figure 8.3. Incoming packer trucks discharge refuse into the receiving area. Refuse is conveyed from receiving area directly into shredders. The shredded refuse is air classified at the shredder outlet. High density material such as shredded metal and pulverized glass are separated out and mechanically conveyed to a disposal bin while light materials, predominantly shredded paper and plastics, are pneumatically conveyed to the shredded refuse storage container.

Refuse leaving storage enters the rotary dryer where the combustion heating value of the material is increased by tumbling it through a stream of hot gas by-passed from the turbine exhaust. Odors emanating from both the storage area and the drying process are eliminated by ducting the drying gases and the pneumatic conveying air into the gas turbine air inlet to be subsequently incinerated by passing through the combustion process.

The shredded and classified refuse is next fed into the fluid bed combustors by the high pressure feeders. The primary function of the feeder and its associated feeder conveyor is to meter the refuse at a rate suitable to control the gas turbine power output. The feeder also serves as an air-lock to pass the material from ambient pressure into the high pressure atmosphere of the combustors while preventing an excessive back flow of combustor gases.

Refuse is pneumatically conveyed from the high pressure feeders into a fluidized bed combustor. In the fluid bed, inert sand-sized particles are buoyed and mixed by an upward flow of air coming from the compressor. Combustion of the refuse maintains the fluid bed particles and combustion products at a temperature between $1500°$ to $1800°F$. The hot fluid bed both retains the incoming refuse and assures rapid and complete combustion.

The exhaust stream from the fluid bed is first passed through an alumina removal chamber. While most of the aluminum is removed by air classification, a small fraction enters the combustor in the form of can and foil fragments, etc. Pure aluminum is molten at combustor temperatures, but particles may be expected to exist within frozen oxide shells in the exhaust. High velocity impingement of such particles on surfaces have led to the development of deposits in the exhaust system during subscale tests. The alumina removal system will be incorporated into the pilot plant to minimize this problem.

From the alumina removal chamber, the hot combustion gases are directed to inertial separators where particulate matter is removed prior to the gases flowing through the turbine and out the exhaust stack.

AMERICAN THERMOGEN HIGH TEMPERATURE PROCESS

The American Thermogen process being developed in a pilot plant would recover (based on a 1,650 tons/day plant) 680,000 pounds/hour of steam (gross), or 320,000 pounds/hour of steam (net); in addition to 410 tons/day of frit.

142   Energy from Solid Waste

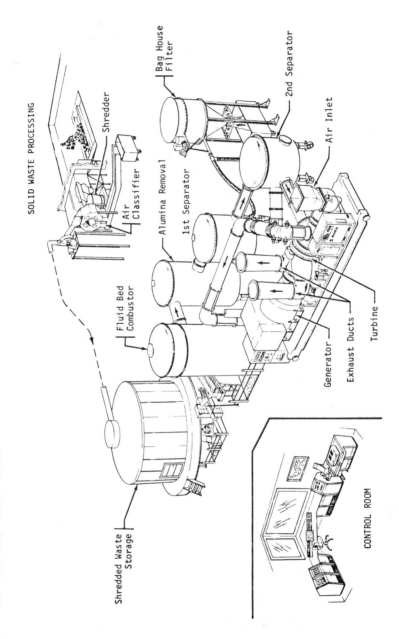

FIGURE 8.3: CPU-400 PILOT PLANT

FIGURE 8.4a: AMERICAN THERMOGEN DESTRUCTOR

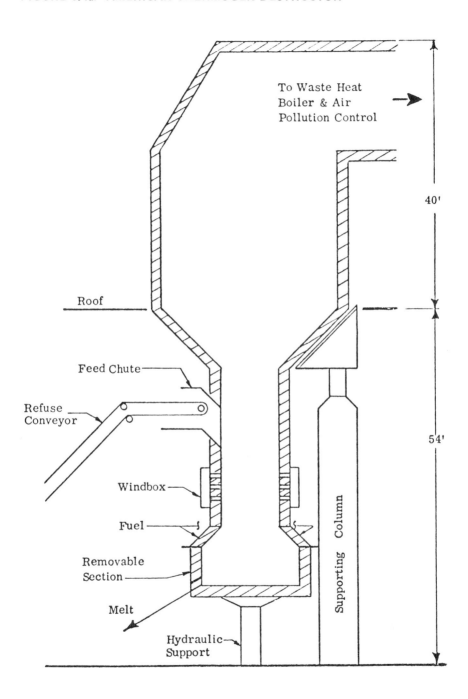

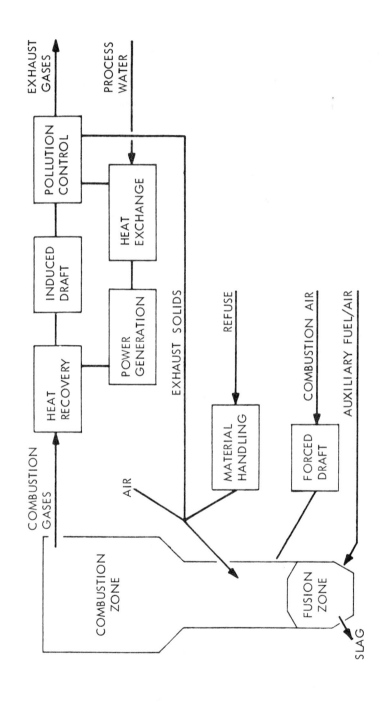

FIGURE 8.4b: AMERICAN THERMOGEN HIGH TEMPERATURE INCINERATOR

Process Descriptions

Nonsegregated, as-received municipal waste is reclaimed from storage pits on a "first in-first out basis" by a positive action conveying system. The conveyor weigh scale records and controls the rate of flow, which varies considerably, depending on the composition of the waste, to maintain a constant total heat release in the destructor. The maintenance of a constant heat release leads to stable system operation, i.e., steam generation and consumption and power generation and consumption.

All openings and transfer points in the materials handling system are sized and designed to obviate any hangups or blockages arising from the completely heterogeneous nature of the waste.

The destructor, shown in Figure 8.4a, reduces the volume of the refuse feed to about 3 to 5% of its original volume by complete burning of all combustible materials, largely in suspension, and by the melting down of all metal and glass objects. The melt-down is accomplished at temperatures of about 3000°F. by burning a relatively small amount of an auxiliary fuel, either oil or gas.

The molten materials are drawn continuously out of the bottom of the destructor through one or more melt taps. The melt stream drops into water in a quench tank where its temperature is dropped precipitously to about 150°F., causing it to shatter through thermal stresses to form a solid, granular, lava-like material. The solidified melt, known as frit, is continuously removed from the quench tank by drag conveyor and is moved to storage bins from where it may be reclaimed for final disposition. This frit has demonstrated value and does not constitute an economic penalty to the process.

Combustion in the melt zone is conducted with less than stoichiometric air to inhibit the formation of NO_x. Secondary air, less than theoretical, is admitted to the destructor just above the melt zone to burn out the combustible portion of basically noncombustible items. Some thermal degradation of organic refuse occurs, and the gaseous products of this pyrolysis join with the combustion products from the melt zone and the burnout zone. A small amount of tertiary air is admitted through the charging chute along with the refuse. This is controlled in amount by controlling the negative pressure just inside the vessel.

A large portion of the refuse is levitated, as it enters the vessel, by the stream of hot gases rising from the lower section. This refuse is burned with an insufficiency of air, admitted tangentially, in the upper expanded section of the vessel with some attendant pyrolysis of organics.

The flue gases containing some gaseous combustible leave the destructor through an overhead duct at temperatures ranging from 1800° to 2000° F. depending on the composition of the waste. Combustion is completed in a waste heat boiler, generating steam and cooling the gases. This steam is used to run turbines which drive plant equipment directly or which generate electricity. Excess steam over and above that required to make the plant self-sufficient, is available for export. All steam generated constitutes a recovered resource.

The flue gases leaving the boiler at 650°F. are water-quenched to near saturation prior to entering a wet scrubbing system where particulates and noxious gases are removed. The wet gases then pass through a separator to remove entrained water droplets before being drawn into the induced draft fan and out the stack. In order to suppress the plume in the stack effluent, the flue gases are mixed with hot air from the air-cooled steam condenser. This exchanger condenses all steam that is not exported from the plant. Water from the frit quencher is used to quench the flue gases. All makeup water for the process, with the exception of boiler feed water makeup, is fed into the frit quencher.

The scrubber water is recirculated and the solids content held at 10% while the net fly ash into the system is removed by filtering a slip stream. The filtrate is returned to the system while the filter cake is returned to the fusion zone of the destructor for incorporation into the melt. The pH and the dissolved solids concentration of the scrubber water are controlled by chemical addition and blowdown, respectively.

Municipal water is treated in various ways depending on its ion content to produce boiler feed water for the waste heat boilers. The blowdown from the boiler and the scrubber is treated before being discharged, thus allowing the entire facility to meet all applicable pollution standards.

Air and water pollutants generated by the system should be minimal. A wet scrubber will be used to control particulate and gaseous emissions from incinerator. Nitrogen oxide emissions may be a problem. Provisions have been made to test blowdown from the boiler and scrubber before discharge from plant. A flow diagram is shown in Figure 8.4b.

TORRAX HIGH TEMPERATURE INCINERATOR

Torrax Systems, Inc. is operating a 75 ton/day demonstration plant which, if scaled up to a 300 tons/day of solid waste input, would generate 45,000 to 70,000 pounds/hour of steam and have as by-products 18 to 24 tons/day of metal and 45 to 60 tons/day of slag.

The Torrax System for high temperature solid waste disposal uses very high temperature preheated combustion air provided by a Super Blast Heater as in Figure 8.5.

Combustion air is filtered (equipment not shown) and then heated in the Super Blast Heater by passing it through silicon carbide tubes around which flow the hot combustion products of a fuel, ordinarily gas or oil. The air is heated to any desired temperature up to 2000°F. and the temperature of the air can be increased or decreased rapidly in order to accommodate changes in demand.

During operation of the system, refuse is charged periodically by conventional means into the top of the Gasifier, and its level is maintained within prescribed limits. As they slowly descend in the Gasifier, most of the readily combustible materials never reach the high temperature zone at the bottom; hot gases permeating up through the refuse decompose the organic materials to form combustible gases.

The material which reaches the bottom is comprised of difficult-to-burn objects and noncombustibles. These materials are burned or liquefied to form a complex silicate slag and a mixture of molten metals. Temperatures at the base of the Gasifier are in the range of 2600° to 3000°F., depending on the refractoriness of the noncombustibles. A liquid mixture of inorganic and metallic materials flows from the Gasifier into a chamber filled with water where an aggregate-quality frit is formed from the inorganic slag and the metal is frozen into small droplets.

Process Descriptions 147

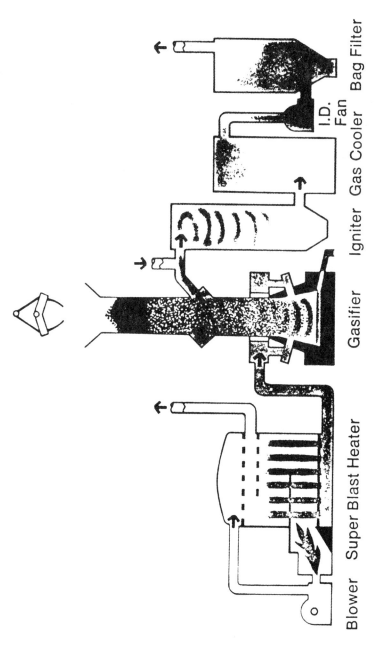

FIGURE 8.5: TORRAX SOLID WASTE DISPOSAL SYSTEM

The gases flowing from the Gasifier contain no free oxygen and consist mainly of carbon monoxide, hydrocarbon gases, and nitrogen. Entrained in this gas stream are particles of carbon, fly ash, etc. This combustible gas-solid mixture is reacted in the Igniter with ambient air. It is estimated that less than 15% excess air is required to carry the combustion reactions to completion. The Igniter operates above 2000°F., and the entrance gas steam is admitted tangentially in order to fuse noncombustible material on the refractory wall of the Igniter. An inorganic slag flows from the base of the Igniter into a water quench tank similar in function and operation to the tank associated with the Gasifier.

The remaining equipment in the process serves to extract heat from the gases issuing from the Igniter and to cleanse the gas stream of particulate matter before releasing it to atmosphere. The last component in the System is a special fabric filter which provides highly efficient removal of particulate contaminants.

CHICAGO NORTHWEST INCINERATOR

This incinerator developed by the Ovitron Corporation, IBW-Martin Incinerator Group, has been in operation since March 1971. The 1,600 ton/day plant produces 440,000 pounds/hour of steam, and 85 tons/day of metals.

The incinerator design for the Chicago Northwest Incinerator (Figure 8.6) is based on the Martin Incinerator System, widely used in Europe. This system with its unique reverse-reciprocating grate is presently being used to burn over 20,000 tons/day of refuse in over 20 installations.

The refuse is taken from the large storage pit and fed directly to the incinerator feed hopper. Directly below the feed hopper is a steeply inclined feed chute which is water cooled to prevent overheating in case of accidental back burning. At the top of the feed chute is a hydraulically operated shutoff gate which is used only during system shutdown to prevent air infiltration into the system.

After the refuse has been stacked in the feed chute, it is fed automatically onto the stoker by means of hydraulic feed ram controlled by the pressure in the under-grate air plenums. The ram is divided into three sections each approximately 7 feet wide, each driven by its own hydraulic cylinder.

The heart of the incinerator is the especially designed reverse-reciprocating stoker. This stoker was specifically designed for burning municipal refuse. The stoker section is inclined at an angle of 26° developing a normal downward flow of the refuse over the grate. Meanwhile the reverse-acting grate bars push the refuse back up the inclined slope, creating a tumbling and mixing action. High intensity air is introduced between the finely ground grate bars, producing a torch-like effect on the refuse. This action, in conjunction with the constant tumbling and turning motion of the refuse, are developed to provide combustion conditions which result in maximum burnout in the shortest length of grates.

The grate bars, constructed of high-grade chrome steel alloy, are capable of withstanding temperatures of over 1500°F. The life of the grate bars is extended through a unique design feature which allows the high-pressure underfire air to cool the grate bars as it enters the combustion chamber. In addition, the expansion and contraction caused by temperature variation is compensated for by expansion joints that separate the stoker sections. The expansion bars are designed to keep the spaces between the various grate bars equal regardless of temperature changes.

To insure long life the driving mechanism for the stoker is continuously lubricated by an oil mist system which is located beneath the stoker in the area cooled by forced draft air.

Although the spacing between the grate bars comprises less than 2% of the total grate area, it is still possible for small siftings or ashes to find their way through the grate. This fine ash is handled by the automatic siftings discharge which extends underneath the air plenum chambers serving the stoker. At regular intervals, high pressure air is directed through the siftings channel, driving the siftings into the ash discharger.

In order to obtain maximum burn-out, the depth of the refuse bed is controlled by the automatic discharge roller located at the end of the grate. As the residue reaches this point, it is dumped into the ash discharger where it is immediately quenched in water. Following quenching, the residue is hydraulically pushed up an inclined slope and allowed to drain. This feature produces a residue with less than 15% moisture, and permits "dry-type" conveying. In addition to quenching the residue, which has temperatures up to 500° or 600°F., the ash discharger also serves as a water seal for the furnace. This seal prevents infiltration of air into the furnace which is under negative pressure.

Supplementing the grate action, the Martin Incinerator is designed to make use of the high-intensity flames created in combustion. Flames and gases are directed up over the stoker area toward the incoming refuse, where they aid in drying and igniting the new refuse. The furnace is designed with a narrow throat area at the upper end of the stoker, where additional high intensity air is introduced through overfire air nozzles. The turbulent action of air and flame in this area produces temperatures exceeding 2000°F., preventing any unburned gases or oxidizable odors from escaping into the boiler and being carried back through to the stack.

The boiler, approximately 40 feet in height, is constructed of membrane water-walled tubes with extruded fins. These various membrane panels are welded together to provide a gas-type enclosure, eliminating problems such as casing corrosion as a result of corrosive gases entering into the area between the water-wall and casing. The use of refractory is limited to an area approximately 15 feet above the grate.

This limited amount of refractory prevents corrosion to the water-walls resulting from various volatile gases produced by the burning of plastics. Burning of plastics can cause corrosion if the temperature is allowed to drop below the dew point temperature of the gases. Since the amount of refractory material used is kept to a minimum, repair and replacement of refractory is also minimal.

The boiler is designed with five passes, providing for a maximum amount of heat recovery, and providing for the collection of a large amount of fly ash in the hoppers directly underneath each pass. The fly ash collected in these hoppers is automatically returned to either the stoker or the ash discharger, where it is mixed with other residue and removed from the plant.

Once the gases have passed through the boiler they are directed through the economizer section to obtain additional heat recovery. An additional fly ash hopper is located beneath the economizer section.

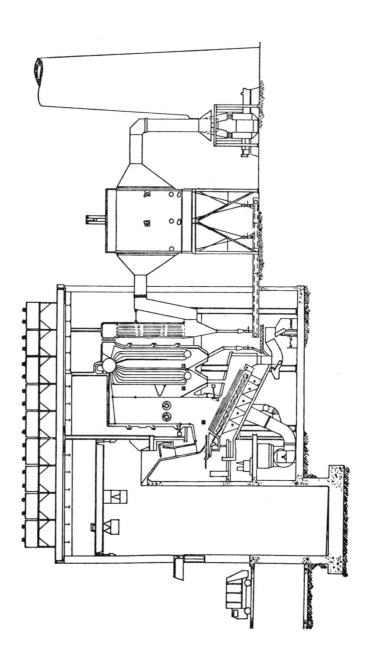

FIGURE 8.6: CHICAGO NORTHWEST INCINERATOR

The convection section of the boiler, consisting of three passes, is designed with widely spaced, in-line tubes to reduce fly ash deposits. This design permits the use of sootblowers to provide automatic, sequential cleaning of the boiler passes. Hence, the incinerator can be operated continuously 24 hours a day, 7 days a week.

Each incinerator system handles 400 tons of refuse per day, and produces approximately 110,000 pounds of steam per hour. A portion of the steam will be used for in-plant use to drive turbines for pump and blower operation. The balance of the steam, less blowdown requirements, is available for sale to outside customers. If at any time there is no immediate use for the excess steam, air-cooled condensers located on the roof can condense 100% of the steam produced.

Incorporated in the side walls of the first boiler pass are two auxiliary burners. These burners are located approximately 20 feet above the grate area and are used to produce steam only when municipal refuse is unavailable for burning, or if the moisture content of the refuse is exceedingly high during periods of adverse weather.

After the combustion gases have passed through the boiler and economizer they are reduced to a temperature of approximately 450°F. The gases are then introduced into an electrostatic precipitator which collects the particulate matter still contained in the gas stream, providing a final particulate concentration of 0.05 grains/cubic foot (0.093 pounds/1,000 pounds) of gas emitted from the stack. To obtain this high collection efficiency, the gas velocity in the precipitators is reduced to less than 3 feet /second.

The specially designed precipitators provide approximately 22,000 square feet of collecting surface and are basically constructed of corten steel which resists corrosion action. Once the dust particles have collected on the collector plates, as a result of the electrical field set-up in the gas passages, the plates are cleaned automatically by a regular rapping action. The fly ash drops into hoppers located beneath the precipitators and is conveyed back to the ash discharger. Before entering the ash discharger, the fly ash is sufficiently moistened in a separate conditioning screw conveyor to prevent dust as the fly ash drops into the discharger.

MONTREAL INCINERATOR

The Montreal incinerator using the Von Roll process, has an input of 1,200 tons/day and an output of 100,000 pounds/hour of steam.

The grate itself consists of three distinct sections: (1) a predrying grate, (2) the main incineration area, and (3) a finishing grate. The two last sections are provided with moving blades breaking the firebed and improving the control of the combustion process.

The first section (Figure 8.7) serves the double purpose of drying the refuse and igniting it, while exposing it to the radiation of the full flow of high-temperature gases, before these enter the boiler.

On the second and third grate sections, the refuse is burned out at a satisfactory rate. The components of these grates have been steadily improved in design and material. The stoker works with only one combustion air fan, an important advantage from the standpoint of operation and maintenance. The

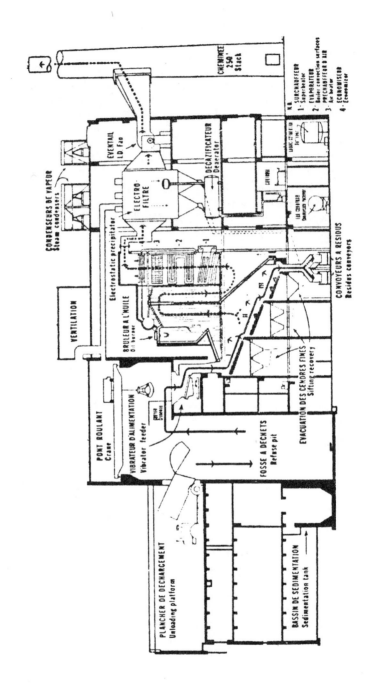

FIGURE 8.7: CROSS-SECTION OF MONTREAL'S STEAM-PRODUCING INCINERATOR

quality of the burnout is determined by the amount of organic matter left in the clinker.

According to German experience, a maximum of 0.3 to 0.5% of fermentable matter indicates a sterile clinker, which after it has been dumped, will not smell nor attract rodents or insects, nor pollute groundwater. The clinker falls, by gravity, through a bifurcated chute, into one of the two transversal conveyors. At the end of each conveyor, a rotating screen separates the scrap metal in the ash. Ash and scrap metal are loaded on trucks. The volume of this residue of incineration is approximately 10 to 15% of the initial volume.

Depending upon the season, the air for combustion may be taken from the outside, from the boiler room and from the garbage pit, always kept under a mild suction to eliminate odors. It is heated to approximately 300°F. in a steam and a flue gas heater before its admission under the grates.

The boiler, a one-drum Eckrohr unit, similar to the one in Frankfurt-am-Main, includes a radiant section, a convection section, a superheater for a mild 100°F. of superheat, an air heater and an economizer. It is built over a cleanout hopper. In the front part of the boiler is a water-cooled combustion chamber with an oil burner for Bunker C fuel, firing downward. It is important that the oil flame remains separate from that part of the combustion chamber in which the refuse is burned: heat transfer conditions and temperature of the two types of flames are quite different and would create undesirable results if generated together.

Therefore, it is only the gases from the two combustion chambers, oil and refuse, that are mixed and delivered to the boiler proper. The refuse combustion chamber is of ample volume to maintain a satisfactory temperature, without excess heat that might cause the refuse cake to melt or the walls to suffer under heavy load conditions. Combustion-chamber temperatures must not exceed 1900°F.

The lower part of the furnace has no water walls: the brickwork consists of silica carbide brick for the part of the walls in direct contact with the firebed. Higher up, the sidewalls are water-cooled.

A Research-Cottrell electrostatic precipitator will remove over 95% of the dust entrained by the gases. Each of the four precipitators handles 100,000 cubic feet per minute of 482°F. gas on a continuous basis. It will reduce the dust to about 0.17 pound/1,000 pounds of gas.

A number of soot blowers will be installed at proper locations. The two chimneys are each 250 feet high.

Although the Montreal Von Roll incinerator will be the first on the North American continent, the design has been widely used since 1954. It is the epitome of a type which includes successive refinements in the art of burning refuse and generating steam simultaneously. Among the largest installations in operation or under construction are the units in Hamburg, The Hague, Tokyo, Frankfurt-am-Main, Vienna, Nuremberg, and Basel, each showing some improvement over the previous one.

USBM HYDROGENATION PROCESS

The U.S. Bureau of Mines has developed a process in a bench-scale pilot plant for processing 100 to 500 gallons/hour of a waste slurry, to recover 1 to 2

154   Energy from Solid Waste

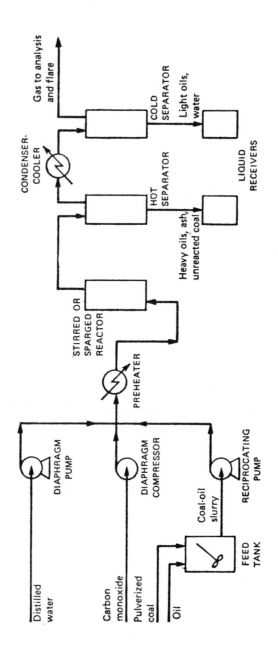

FIGURE 8.8: BUREAU OF MINES HYDROGENATION PROCESS

barrels of low-sulfur oil per ton of refuse, having a fuel value of 13,000 to 16,000 Btu/pound.

Both batch and continuous reactor systems have been utilized. Batch studies were conducted in 0.5 liter and 1 liter stainless steel autoclaves. White pine wood chips and newsprint were used as source of crude cellulose.

Figure 8.8 presents a simplified schematic diagram of the continuous bench-scale unit for converting waste to oil by carbon monoxide-water treatment. The system was designed to operate at maximum conditions of 5,000 psig and 500°C., with feed rates of 100 to 500 grams/hour of waste slurry and 10 standard cubic feet per hour of carbon monoxide.

The combined stream of carbon monoxide and liquid feed was preheated under pressure and injected into the bottom of the heated reactor. The liquid and gas exited from the top of the reactor and separated in the high-pressure recovery system. The liquid collected was intermittently discharged into secondary receivers at atmospheric pressure, while gas was continuously released through a back-pressure regulator.

GARRETT PYROLYSIS PROCESS

The pyrolysis process developed by Garrett Research and Development Company has been investigated in a pilot plant, and will be further tested in a demonstration plant. A commercial plant based on this process could deliver 480 tons/day of oil based on 2,000 tons/day input of solid wastes.

The Garrett process shown in Figure 8.9 is designed to recover salable heating fuels, glass and magnetic metals. The organic portion of these wastes is converted to low sulfur oil, char and gas using a flash pyrolysis process. The process is designed to be expanded into an integrated series of processing stages for the recovery of over 90% of the raw materials contained in municipal refuse. Incoming solid wastes are shredded, dried and passed through an air classifier which separates most of the metals, glass, and other inorganic materials. The overhead stream from the air classifier is then subjected to a two-stage screening to improve separation of inorganics. The remaining refuse is shredded a second time and then pyrolyzed, where it is broken down into smaller molecules through the application of heat in the absence of oxygen.

Laboratory studies of the pyrolysis process resulted in the production of approximately one barrel of good quality oil per ton of as-received refuse. Such refuse usually will also yield about 140 pounds of magnetic metals, 120 pounds of glass, and 160 pounds of char.

The flash pyrolysis operation which is the heart of the Garrett process was researched for over a year in a continuous laboratory reactor. Product yields, quality, and the initial favorable economic projections have since been confirmed at a 4 ton/day pilot plant during an 18-month period of operations at LaVerne, California.

Small quantities of water produced in the process are discharged as liquid. The water contains only small quantities of dissolved solids, but is heavily loaded with organics and thus has a high BOD. Sewer acceptance tests show that there are no water disposal problems, provided secondary sewage treatment facilities are available.

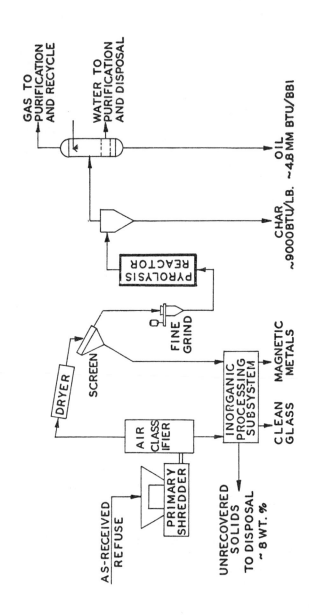

FIGURE 8.9: GARRETT PYROLYSIS PROCESS

Process Descriptions

The quantity of solid material going to landfill from a 100 ton/day plant is only 16 tons/day. The solid debris contains unrecovered glass, aluminum, copper, zinc, nickel and other sterile material. GR&D is currently investigating the economics of reclamation of nonferrous metals, and if a system can be developed, the tonnage of solid rejects going to landfill will be halved.

UNION CARBIDE OXYGEN REFUSE CONVERTER SYSTEM

Union Carbide has had a 5 ton/day pilot plant in operation since early 1971 to produce fuel gas and slag.

A plant with a daily capacity of 1,000 tons of mixed municipal refuse (MMR) would consist of three shaft furnaces and gas cleaning trains with a rating of 334 tons/day. These furnaces accept MMR up to 4 feet in each dimension. Larger items such as sofas, tables, etc., require shearing or shredding. The gas cleaning train consists of an electrostatic precipitator, acid absorber and a condenser. The arrangement of these components is shown in Figure 8.10.

In operation MMR is fed from packer trucks into the furnaces. Gaseous oxygen is fed continuously into the bottom of the furnaces. The molten slag formed by the high temperature in the bottom of the furnaces is tapped continuously and quenched in water.

The gas produced in the furnace from the combustible portion of the refuse exits at low temperature near the top of the furnace. This gas contains significant moisture, some oils and fly ash. The fly ash content is about 0.08 grains/cubic foot (corrected to 12% CO_2). The volume of this gas is only 5 to 10% of the gas volume produced in a conventional incinerator due to the absence of nitrogen and excess air. Production of nitric oxides is also virtually precluded for the same reason.

The exit gas from the furnace passes through an electrostatic precipitator to remove condensed droplets of oil and the bulk of the remaining fly ash. The oil and fly ash are recycled back to the furnace where the oils are cracked to gaseous products and the fly ash exits with the molten residue.

The gases then pass into an acid absorber where a neutralizing solution removes HCl, H_2S and organic acids. The aqueous solution of salts is continuously bled from the recycled absorber liquid, and fed to the furnace where the salts are eliminated with the slag. Moisture is removed from the saturated gas stream in a condenser.

Upon completion of the cleaning step, the gas contains about 0.008 grains/cubic foot of dry particulate matter (corrected to 12% CO_2). The fuel gas, although its heating value is only about one-third of natural gas, does have a flame temperature and heat transfer characteristics similar to natural gas. It can be used as a substitute fuel for natural gas, oil or coal in utility boilers or other large heat consuming operations. Patents are pending.

MONSANTO LANDGARD SYSTEM

Monsanto Enviro-Chem has operated a 35 ton/day pilot plant for the past few years to produce pyrolysis gas.

The Landgard system is designed to be a totally self-contained operation. The system is based on pyrolysis with the primary objective being the disposal of all

158   Energy from Solid Waste

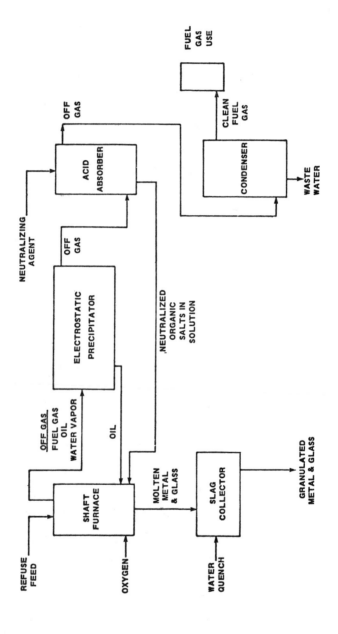

FIGURE 8.10: UNION CARBIDE OXYGEN REFUSE CONVERTER

FIGURE 8.11: MONSANTO LANDGARD SYSTEM

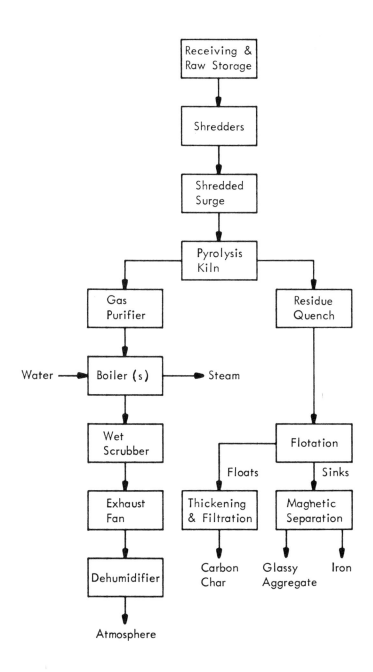

types of solid waste while offering practical opportunities for resource recovery and recycling. Figure 8.11 presents a schematic diagram of a commercial plant. The Enviro-Chem Landgard System encompasses all operations for receiving, handling, shredding and pyrolyzing waste; for quenching and separating the residue; for generating steam from waste heat; and for purifying the off-gases.

Waste will be received in trucks, and is fed into either of two shredder lines directly from the trucks or from the floor by use of front-end loader. The plant layout will allow for the future installation of a third shredder for standby service. After shredding, waste is conveyed to a shredded waste storage system. From the storage system, shredded waste is continuously fed into the rotary kiln.

Pyrolysis of shredded waste occurs in a refractory lined rotary kiln. Shredded waste feed and direct-fire fuel enter opposite ends of the kiln. Heat is supplied by firing fuel oil. Countercurrent flow of gases and solids expose the feed to progressively higher temperatures as it passes through the kiln, so that first drying and then pyrolysis occurs. The finished residue is exposed to the highest temperature just before it is discharged from the kiln.

The kiln is specially designed to uniformly expose solid particles to high temperatures before they are discharged. This maximized the pyrolysis reaction.

The hot residue is discharged from the kiln into a water-filled quench tank. A conveyor elevates the wet residue from the quench tank into a flotation separator. Light material floats off as a carbon char slurry and is thickened and filtered to remove the water. Clarified water and filtrate are recycled for reuse. Carbon char is conveyed to a storage pile prior to truck transport from the site. Heavy material is conveyed from the bottom of the flotation separator to a magnetic separator for removal of iron. Iron is deposited in a storage area or directly into a rail car or truck. The balance of the heavy material, glassy aggregate, passes through screening equipment and then is stored on-site in a pile.

Pyrolysis gases are drawn from the kiln into a refractory lined gas purifier where they are mixed with air and burned. The gas purifier prevents discharge of combustible gases to the atmosphere and subjects them to temperatures high enough for destruction of odors.

Hot combustion gases from the gas purifier pass through water tube boilers where heat is exchanged to produce steam. Exit gases from the boilers are further cooled and cleaned of particulate matter as they pass through a water spray scrubbing tower. In order to minimize liquid flows, the scrubber is operated adiabatically. Provisions are included to allow gas purifier exit gases to completely or partially bypass the boilers and enter the scrubber tower directly.

Scrubbed gases then enter an induced draft fan which provides the motive force for moving the gases through the entire system. Gases exiting the induced draft fan are saturated with water. To suppress formation of a steam plume, the gases are passed through a dehumidifier in which they are cooled (by ambient air) as part of the water is removed and recycled. Cooled process gases are then combined with heated ambient air just prior to discharge from the dehumidifier.

Solids are removed from the scrubber by diverting part of the recirculated water to a thickener. Underflow from the thickener is transferred to the quench tank, while the clarified overflow stream is recycled to the scrubber.

Normally all the water leaving this system will be carried out with the residue or evaporated from the scrubber. An occasional process upset may allow too much water into the system and make it necessary to purge the excess. This purge stream will be discharged to the sanitary sewer at a maximum flow of 75 gallons/minute.

BUREAU OF MINES PYROLYSIS PROCESS

The U.S. Bureau of Mines has developed a pyrolysis process to recover oil, gas, tar, and char.

The plant consists essentially of an electric furnace, cylindrical steel retort, condensing and scrubbing train for product recovery, and gas-metering and sampling devices, as shown in Figure 8.12.

The electric furnace (2) is 26 inches inside diameter and 48 inches deep and is heated by nickel-chromium resistors spaced evenly in the furnace wall. The retort (3) is 18 inches in diameter and 26 inches deep and is made of 16-gauge steel in the wall and 10-gauge steel in top and bottom. Gases and vapors exit from the retort through a 2-inch diameter offtake pipe and enter an air-cooled trap (4), where tar and heavy oils are collected. The gases and vapors are cooled to room temperature in two water-cooled condensers (5) connected in series where additional heavy oil and liquor are collected. Final traces of heavy-oil mist are removed by one of the alternate electrostatic precipitators (6).

The gas then passes successively through packed scrubbers, where ammonia is removed with sulfuric acid (7), and carbon dioxide and hydrogen sulfide are removed with caustic soda solution (9). The scrubbed gases pass to the large (11) and small (14) meters, which are geared together so that 99% of the gas passes through the large meter and is flared. The suction side of the small meter is cross-connected to the inlet side of the large meter so that 1% of the gas passes through a drying tube (12) and a condenser (13) immersed in acetone and solid carbon dioxide, where light oil is removed. Light oil recovered from the gas that passes through the small meter is calculated to the total gas yield.

The gas from the condenser passes through the small meter (14) to the gas holder (15), and representative samples are taken from the holder for analysis. Steam is used to purge the condensers and piping at the conclusion of the test.

BATTELLE PYROLYSIS INCINERATION PROCESS

The Pacific Northwest Laboratories of Battelle have been operating a 10 ton/day pilot plant to produce fuel gas and slag.

The basic pyrolysis-incineration process is performed in a completely closed system that does not release fly ash, fumes, odors, or smoke to the environment. The heart of the process is a closed vertical reactor where the refuse is progressively dried, charred, and finally oxidized at relatively low temperatures under carefully controlled conditions. The refuse undergoes these transformations in a packed bed settling under the force of gravity while reactant and combustion product gases rise countercurrent to the direction of solids movement.

To produce a heating gas, the solid char, the product of pyrolysis in the upper portion of the reactor, is oxidized in the bottom part of the reactor by a mixture

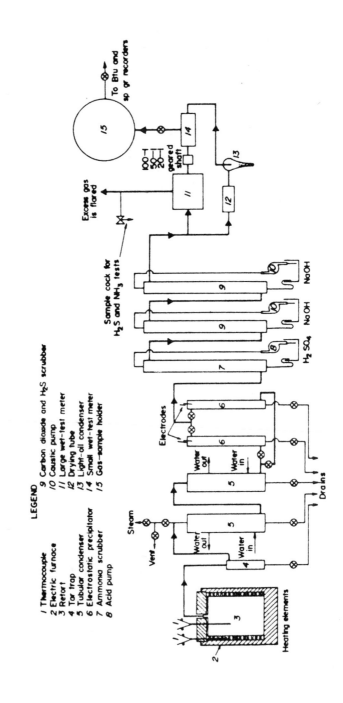

FIGURE 8.12: USBM PILOT PLANT USED TO PYROLYZE MUNICIPAL AND INDUSTRIAL REFUSE

of oxygen (from either air or commercial oxygen) and steam. The hot reaction product gases continue upward and release their heat to cause charring of the entering refuse. Finally, the residual heat in the gases evaporates moisture from the entering refuse at the top of the reactor. The gases which leave the reactor contain hydrogen, oxides of carbon, water vapor, and a mixture of hydrocarbons. These gases may be cleanly burned in a secondary burner since they contain no ashy materials.

Alternately, they may be processed for recovery of organic compounds, further treated to produce a heating gas, or processed still further to yield a 75% hydrogen and carbon monoxide mixture which may be used to synthesize methane. This offgas could also be used to produce electricity.

This processing method may not require either presorting or shredding and will be able to accept difficult-to-handle materials such as partially dewatered sewage sludge, automobile tires, crankcase oils, and certain types of hazardous chemicals.

POLLUTANT REMOVAL HANDBOOK 1973

by Marshall Sittig

The purpose of this handbook is to provide a one volume practical reference book showing specifically how to remove pollutants, particularly those emanating from industrial processes. This book contains substantial technical information.

This volume is designed to save the concerned reader time and money in the search for pertinent information relating to the control of specific pollutants. Through citations from numerous reports and other sources, hundreds of references to books and periodicals are given.

In this manner this book constitutes a ready reference manual to the entire spectrum of pollutant removal technology. While much of this material is presumably available and in the public domain, the locating thereof is a tedious, time-consuming, and expensive process.

The book is addressed to the industrialist, to local air and water pollution control officers, to legislators who are contemplating new and more stringent control measures, to naturalists and conservationists who are interested in exactly what can be done about the effluents of local factories, to concerned citizens, and also to those eager students who can foresee new and brilliant careers in the fields of antipollution engineering and pollution abatement.

During the past few years, the words "pollution", "environment" and "ecology" have come into more and more frequent usage and the cleanliness of the world we live in has become the concern of all people. Pollution, for example, is no longer just a local problem involving litter in the streets or the condition of a nearby beach. Areas of the oceans, far-reaching rivers and the largest lakes are now classified as polluted or subject to polluting conditions. In addition, very surprisingly, lakes and streams remote from industry and population centers have been found to be contaminated.

This handbook therefore gives pertinent and concise information on such widely divergent topics as the removal of oil slicks in oceans to the containing of odors and particulates from paper mills.

Aside from the practical considerations, including teaching you where to look further and what books and journals to consult for additional information, this book is also helpful in explaining the new lingo of pollution abatement, which is developing new concepts and a new terminology all of its own, for instance: "particulates, microns, polyelectrolytes, flocculation, recycling, activated sludge, gas incineration, catalytic conversion, industrial ecology, etc."

In order to have a safe and healthful environment we must all continue to learn and discover more about the new technology of pollution abatement. Every effort has been made in this manual to give specific instructions and to provide helpful information pointing in the right direction on the arduous and costly antipollution road that industry is now forced to take under ecologic and sociologic pressures. The world over, technological and manpower resources are being directed on an increasing scale toward the control and solution of contamination and pollution problems.

In the United States of America we are fortunate in receiving direct help from the numerous surveys together with active research and development programs that are being supported by the Federal Government to help industry and municipalities control their wastes and harmful emissions.

A partial and condensed table of contents is given here. The book contains a total of 128 subject entries arranged in an alphabetical and encyclopedic fashion. The subject name refers to the polluting substance and the text underneath each entry tells how to combat pollution by said substance:

INTRODUCTION
ACIDS
ADIPIC ACID
ALDEHYDES
ALKALIS
ALKALI CYANIDES
ALUMINUM
ALUMINUM CELL EXIT GASES
ALUMINUM CHLORIDE
ALUMINUM SILICATE PIGMENT
AMMONIA
AMMONIUM PHOSPHATE
AMMONIUM SULFATE
AMMONIUM SULFIDE
AMMONIUM SULFITE
AROMATIC ACIDS & ANHYDRIDES
ARSENIC
ASBESTOS
AUTOMOTIVE EXHAUST EFFLUENTS
BARIUM
BERYLLIUM
BLAST FURNACE EMISSIONS
BORON
BREWERY WASTES
CADMIUM
CARBON BLACK

- CARBON MONOXIDE
- CARBONYL SULFIDE
- CEMENT KILN DUSTS
- CHLORIDES
- CHLORINATED HYDROCARBONS
- CHLORINE
- CHROMIUM
- CLAY
- COKE OVEN EFFLUENTS
- COLOR PHOTOGRAPHY EFFLUENTS
- COPPER
- CRACKING CATALYSTS
- CYANIDES
- CYCLOHEXANE OXIDATION WASTES
- DETERGENTS
- DYESTUFFS
- FATS
- FERTILIZER PLANT EFFLUENTS
- FLOUR
- FLUORINE COMPOUNDS
- FLY ASH
- FORMALDEHYDE
- FOUNDRY EFFLUENTS
- FRUIT PROCESSING INDUSTRY EFFLUENTS
- GLYCOLS
- GREASE
- HYDRAZINE
- HYDROCARBONS
- HYDROGEN CHLORIDE
- HYDROGEN CYANIDE
- HYDROGEN FLUORIDE
- HYDROGEN SULFIDE
- IODINE
- IRON
- IRON OXIDES
- LAUNDRY WASTES
- LEAD
- LEAD TETRAALKYLS
- MAGNESIUM CHLORIDE
- MANGANESE
- MEAT PACKING FUMES
- MERCAPTANS
- MERCURY
- METAL CARBONYLS
- MINE DRAINAGE WATERS
- NAPHTHOQUINONES
- NICKEL
- NITRATES
- NITRITES
- NITROANILINES
- NITROGEN OXIDES
- OIL
- OIL (INDUSTRIAL WASTE)
- OIL (PETROCHEMICAL WASTE)
- OIL (PRODUCTION WASTE)
- OIL (REFINERY WASTE)
- OIL (TRANSPORT SPILLS)
- OIL (VEGETABLE)
- ORGANIC VAPORS
- OXYDEHYDROGENATION PROCESS EFFLUENTS
- PAINT AND PAINTING EFFLUENTS
- PAPER MILL EFFLUENTS
- PARTICULATES
- PESTICIDES
- PHENOLS
- PHOSGENE
- PHOSPHATES
- PHOSPHORIC ACID
- PHOSPHORUS
- PICKLING CHEMICALS
- PLASTIC WASTES
- PLATING CHEMICALS
- PLATINUM
- PROTEINS
- RADIOACTIVE MATERIAL
- RARE EARTH
- ROLLING MILL DUST & FUMES
- ROOFING FACTORY WASTES
- RUBBER
- SELENIUM
- SILVER
- SODA ASH
- SODIUM MONOXIDE
- SOLVENTS
- STARCH
- STEEL MILL CONVERTER EMISSIONS
- STRONTIUM
- SULFIDES
- SULFUR
- SULFUR DIOXIDE
- SULFURIC ACID
- TANTALUM
- TELLURIUM HEXAFLUORIDE
- TETRABROMOETHANE
- TEXTILE INDUSTRY EFFLUENTS
- THIOSULFATES
- TIN
- TITANIUM
- TRIARYLPHOSPHATES
- URANIUM
- VANADIUM
- VEGETABLE PROCESSING INDUSTRY EFFLUENTS
- VIRUSES
- ZINC

ISBN 0-8155-0489-6